21世纪高等学校计算机规划教材

21st VBentury University Planned Textbooks of VBomputer SVBienVBe

VB语言程序设计实验教程

VB Programming Experiments Tutorial

史晓峰 刘超 主编
薄海玲 王莉莉 副主编

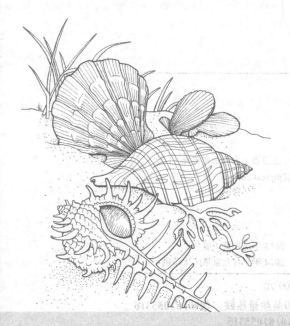

高校系列

人民邮电出版社

北京

图书在版编目（CIP）数据

VB语言程序设计实验教程 / 史晓峰, 刘超主编. --
北京：人民邮电出版社，2015.2
　21世纪高等学校计算机规划教材. 高校系列
　ISBN 978-7-115-32805-2

Ⅰ. ①V… Ⅱ. ①史… ②刘… Ⅲ. ①BASIC语言—程序设计—高等学校—教材 Ⅳ. ①TP312

中国版本图书馆CIP数据核字（2015）第030527号

内 容 提 要

本书是《VB语言程序设计教程（第2版）》一书的配套实验教程。全书共14章，前10章讲解程序设计实验内容与习题，包括主教材对应的知识点、实验内容及测试题，第11章介绍综合性实验，第12章和第13章主要介绍 Visual Basic 课程设计题目和开发实例，第14章为自测综合练习题。

本书既可作为高等院校 Visual Basic 程序设计课程的辅导教材，也适合单独作为学习 Visual Basic 语言的参考用书。

◆ 主　编　史晓峰　刘　超
　副主编　薄海玲　王莉莉
　责任编辑　武恩玉
　责任印制　沈　蓉　彭志环

◆ 人民邮电出版社出版发行　北京市丰台区成寿寺路11号
　邮编 100164　电子邮件 315@ptpress.com.cn
　网址 https://www.ptpress.com.cn
　北京盛通印刷股份有限公司印刷

◆ 开本：787×1092　1/16
　印张：12.75　　　　　　　2015年2月第1版
　字数：331千字　　　　　　2024年8月北京第16次印刷

定价：32.00元

读者服务热线：(010)81055256　印装质量热线：(010)81055316
反盗版热线：(010)81055315
广告经营许可证：京东市监广登字 20170147 号

前　言

Visual Basic 程序设计语言是一门实践性很强的课程。要学好它，必须做大量的上机实验和习题，让学生动手调试程序和编写程序，进而让学生建立编程思维。为此，我们编写了这本《VB 语言程序设计实验教程》。本书可单独使用，也可与杨忠宝等编写的《VB 语言程序设计教程（第 2 版）》一书配合使用。

本书按照 Visual Basic 程序设计的特点，采用"任务驱动"方式，精心设计每一个实验内容。其中，前两章以实验指导的形式给出实验内容，指导学生一步一步完成操作，逐渐掌握 Visual Basic 开发环境的使用和程序设计的步骤。全书各章实验题目由浅入深，理论联系实际，注重典型算法的选择。实验题目和测试题的题型与全国计算机等级考试题型一致，题目具有较强的针对性和实践性。通过实验和测试题目可使读者掌握 Visual Basic 程序设计与调试方法，巩固所学知识，培养实际编程能力。本书非常适合初学者使用。

本书主要包括以下内容。

（1）知识点。知识点的形式简洁明了，不仅概括了各章应掌握的主要内容，也概括了教学的重点内容，能使学生快速复习并提高学习效率。同时，也为读者查找命令提供了很大的方便。

（2）实验内容。各章实验的详细程度各不相同，第 1 章、第 2 章实验的操作步骤相当详细；第 3 章～第 10 章，各章只列出实验内容。实验内容包括 3 种题型：程序改错、程序填空和编程题，内容均针对典型算法，由浅入深；第 11 章是综合性实验，只给出一些参考题目，读者可综合运用前面所学知识完成一些小型系统的开发，真正体会到用 Visual Basic 解决实际问题的乐趣；第 12 章和第 13 章为课程设计内容，读者可运用 Visual Basic 开发数据库管理系统等应用程序，使所学的知识得到升华，与办公和专业充分结合。

（3）测试题。精选各类客观和主观题目，测试题覆盖教学的主要内容，设计时考虑到既要全面又避免重复，尤其突出了教学的重点内容。同时，选择题目时也针对全国计算机等级考试的特点，题目难度、类型和内容均与其相近。

本书的所有实验题目和测试题中涉及的程序都已在 Visual Basic 6.0 中文企业版中调试通过。由于解决一个问题通常可采用不同方法、设计不同风格的程序界面、编写出多个程序，而本书中有些题目没有给出程序界面和源程序，读者可按自己的思路编写出不同的、体现个人风格的程序。若需要程序答案，请与编者联系。

本书由长期工作在教学一线并具有丰富程序设计教学经验的多位教师共同编写而成。全书由史晓峰、刘超担任主编，薄海玲、王莉莉担任副主编。史晓峰编写第 1 章、第 2 章、第 4 章和第 10 章、第 11 章，刘超编写第 6 章、第 8 章、第 12 章、第 13 章，薄海玲编写第 3 章、第 9 章和第 5 章"过程与函数"部分的实验题目和测试题，王莉莉编写第 7 章、第 14 章和第 5 章的"数组"部分的实验题目和测试题。感谢长春工程学院计算机基础教学中心对编写本书的大力支持。

由于时间仓促，书中难免有错误和不足之处，请读者批评指正。E-mail：shixf868@sina.com。

编者

2014 年 11 月

目 录

第1章 Visual Basic 概述 ·············· 1
- 1.1 知识点 ··· 1
  - 1.1.1 Visual Basic 的版本 ················ 1
  - 1.1.2 Visual Basic 的主要特点 ········ 1
  - 1.1.3 Visual Basic 6.0 的启动 ·········· 1
  - 1.1.4 Visual Basic 的退出 ················ 1
  - 1.1.5 Visual Basic 的开发环境 ········ 1
  - 1.1.6 Visual Basic 常用的文件类型 ···· 2
- 1.2 实验内容 ·· 2
  - 1.2.1 认识 Visual Basic 6.0 ············· 2
  - 1.2.2 简单的 Visual Basic 加法程序 ···· 9
- 1.3 测试题 ·· 13

第2章 Visual Basic 应用程序设计过程 ·· 15
- 2.1 知识点 ·· 15
  - 2.1.1 面向对象程序设计的基本概念 ···· 15
  - 2.1.2 窗体 ·· 15
  - 2.1.3 命令按钮 ································ 16
  - 2.1.4 标签 ·· 17
  - 2.1.5 文本框 ···································· 17
  - 2.1.6 程序设计过程 ························ 17
- 2.2 实验内容 ·· 18
  - 2.2.1 单个窗体实验 ························ 18
  - 2.2.2 多个窗体实验 ························ 23
  - 2.2.3 文本框实验 ···························· 26
- 2.3 测试题 ·· 27

第3章 VB 语言基本知识 ·············· 29
- 3.1 知识点 ·· 29
  - 3.1.1 Visual Basic 的数据类型 ······· 29
  - 3.1.2 常量和变量 ···························· 29
  - 3.1.3 运算符与表达式 ···················· 29
  - 3.1.4 常用内部函数 ························ 30
- 3.2 实验内容 ·· 32
- 3.3 测试题 ·· 41

第4章 程序的控制结构 ·············· 46
- 4.1 知识点 ·· 46
  - 4.1.1 顺序结构 ································ 46
  - 4.1.2 选择结构 ································ 47
  - 4.1.3 循环结构 ································ 49
- 4.2 实验内容 ·· 49
  - 4.2.1 顺序结构实验 ························ 49
  - 4.2.2 选择结构实验 ························ 55
  - 4.2.3 循环结构实验 ························ 62
- 4.3 测试题 ·· 67

第5章 数组与过程 ······················ 72
- 5.1 知识点 ·· 72
  - 5.1.1 数组 ·· 72
  - 5.1.2 Sub 子过程 ···························· 73
  - 5.1.3 Function 自定义函数过程 ····· 73
  - 5.1.4 子过程和函数过程的参数传递 ···· 73
  - 5.1.5 变量的作用域与生存期 ········ 74
- 5.2 实验内容 ·· 74
  - 5.2.1 一维数组实验 ························ 74
  - 5.2.2 二维数组实验 ························ 81
  - 5.2.3 Function 过程和 Sub 过程实验 ···· 86
- 5.3 测试题 ·· 91

第6章 Visual Basic 常用控件 ···· 101
- 6.1 知识点 ·· 101
  - 6.1.1 单选按钮控件 ······················ 101
  - 6.1.2 复选框控件 ·························· 101
  - 6.1.3 框架控件 ······························ 101
  - 6.1.4 列表框控件 ·························· 102
  - 6.1.5 组合框控件 ·························· 102
  - 6.1.6 图片框控件 ·························· 103

6.1.7 图像框控件 103
6.1.8 滚动条控件 104
6.1.9 计时器控件 104
6.1.10 直线控件（Line）与形状控件（Shape） 104
6.1.11 文件系统控件 104
6.2 实验内容 105
　6.2.1 单选按钮、复选框、框架、列表框和组合框控件实验 105
　6.2.2 图片框、图像框、计时器和滚动条控件实验 113
　6.2.3 文件系统、直线、形状控件与绘图实验 119
6.3 测试题 124

第 7 章 Visual Basic 高级控件 128
7.1 知识点 128
　7.1.1 公共对话框控件 128
　7.1.2 Windows 公用控件 128
　7.1.3 工具箱中添加"高级控件"选项卡 129
　7.1.4 ActiveX 控件添加到工具箱中 129
7.2 实验内容 129
7.3 测试题 135

第 8 章 菜单 138
8.1 知识点 138
　8.1.1 菜单的组成 138
　8.1.2 菜单编辑器 138
　8.1.3 下拉式菜单 139
　8.1.4 弹出式菜单 139
8.2 实验内容 139
8.3 测试题 147

第 9 章 文件操作 149
9.1 知识点 149
　9.1.1 文件分类 149

9.1.2 顺序文件 149
9.1.3 随机文件 149
9.2 实验内容 150
9.3 测试题 157

第 10 章 数据库应用程序设计 160
10.1 知识点 160
10.2 实验内容 160
　10.2.1 用 Adodc 控件连接数据库和表实验 160
　10.2.2 设计学生信息查询窗体 163
10.3 测试题 165

第 11 章 Visual Basic 综合性实验 167

第 12 章 Visual Basic 课程设计基础 169
12.1 概述 169
12.2 课程设计的要求 169
12.3 课程设计预备知识 170
12.4 课程设计参考题目 170

第 13 章 Visual Basic 课程设计实例 172
13.1 系统总体设计 172
13.2 数据库设计 173
13.3 详细设计 174
13.4 编程调试运行 178

第 14 章 自测综合练习题 184
14.1 综合练习一 184
14.2 综合练习二 189

参考文献 196

第 1 章
Visual Basic 概述

1.1 知识点

1.1.1 Visual Basic 的版本

Microsoft Visual Basic（VB）是在 Windows 操作平台下设计应用程序的最迅速、最简捷的工具之一。VB 提供学习版、专业版和企业版。

1.1.2 Visual Basic 的主要特点

（1）可视化的、友好的集成开发环境。
（2）面向对象的程序设计。
（3）支持多种数据库系统的访问。
（4）支持动态数据交换、动态链接库、对象的链接与嵌入以及 ActiveX 技术。
（5）完备的 Help 联机帮助功能。

1.1.3 Visual Basic 6.0 的启动

（1）选择开始/程序/Microsoft Visual Basic 6.0 中文版/Microsoft Visual Basic 6.0 中文版。
（2）开始/运行：找到 Visual Basic 的执行文件 "VB 6.exe"/确定。
（3）单击桌面快捷方式启动 Visual Basic。

1.1.4 Visual Basic 的退出

（1）单击"文件"菜单栏的"退出"命令。
（2）单击窗口右上角的"关闭"按钮。
（3）按 Alt+Q 组合键退出。
（4）选左上角"控制菜单"中的"关闭"。

1.1.5 Visual Basic 的开发环境

其包括标题栏、工具栏、菜单栏、工具箱、窗体设计器、代码编辑器窗口、工程资源管理器、窗体布局窗口等。

1.1.6　Visual Basic 常用的文件类型

（1）工程文件（.vbp）。
（2）窗体文件（.frm）。
（3）窗体的二进制数据文件（.frx）。
（4）标准模块文件（.bas）。
（5）类模块文件（.cls）。
（6）可执行文件（.exe）。

1.2　实验内容

1.2.1　认识 Visual Basic 6.0

【实验要求】

上机前认真阅读实验内容，上机时按照指导书上的实验步骤独立完成并掌握实验内容。

【实验目的】

1. 掌握 VB 的启动、退出。
2. 掌握 VB 6.0 的集成开发环境及各个组成部分的使用方法。
3. 了解程序设计的基本原理和方法。

【实验步骤】

1．Visual Basic 6.0 的启动

用鼠标单击任务栏中的"开始"按钮，选择"程序"菜单中的"Microsoft Visual Basic 6.0 中文版"程序组中的"Microsoft Visual Basic 6.0 中文版"程序项，如图 1-1 所示。

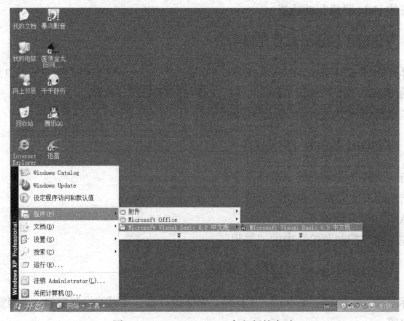

图 1-1　Visual Basic 6.0 中文版的启动

VB 6.0 启动后，屏幕上会首先显示一个"新建工程"对话框，如图 1-2 所示。在弹出的对话框中双击"新建"选项卡中的"标准 EXE"项或直接单击"打开"按钮，即可进入 VB 6.0 集成开发环境，如图 1-3 所示。

图 1-2　"新建工程"对话框

图 1-3　Visual Basic 6.0 集成开发环境

2. Visual Basic 的工作模式

VB 有 3 种工作模式：设计、中断与运行，系统当前处于何种工作模式会在标题栏中显示。

（1）VB 启动后，标题栏中显示"工程 1-Microsoft Visual Basic[设计]"，其中的"设计"说明当前处于"设计模式"，可以进行用户界面的设计和代码的编写。

（2）单击"运行"菜单项中的"启动"或单击工具栏中的"启动"按钮（如图 1-4 所示）或按 F5 键，进入运行状态，此时标题栏中显示"工程 1-Microsoft Visual Basic[运行]"。

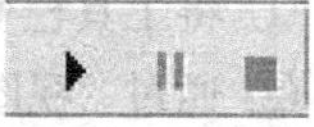

图 1-4　启动按钮

（3）单击"运行"菜单项中的"中断"或单击工具栏中的"中断"按钮或按 Ctrl+Break 组合

键,此时标题栏中显示"工程 1-Microsoft Visual Basic[Break]"。

单击"运行"菜单项中的"结束"或单击工具栏中的"结束"按钮,结束程序运行。

3. 工具栏

VB 6.0 提供了 4 种工具栏:编辑、标准、窗体编辑器和调试。启动 Visual Basic 时默认,只有"标准"工具栏显示在窗体上,其形式及功能如图 1-5 所示。

图 1-5 标准工具栏

(1)工具栏的打开,有两种方法。

① 通过"视图"菜单中的"工具栏"选择。选择"视图"菜单中的"工具栏"命令,系统弹出其下一级子菜单,如图 1-6 所示。单击相应菜单命令即可将相应工具栏显示在窗体上。前面带"√"项目表示该工具栏已显示在窗体上。

② 通过快捷菜单选择。在工具栏的空白处单击鼠标右键,在弹出的快捷菜单中选择,如图 1-7 所示。

图 1-6 工具栏菜单命令　　　　　　图 1-7 快捷方式工具栏

通过以上任一方法选择"窗体编辑器"子命令,则在窗体中将显示窗体编辑器工具栏,如图 1-8 所示。图 1-8 所示的工具栏为浮动形式,有标题栏和关闭按钮。双击标题栏或用鼠标左键拖动其标题移到常用工具栏的下方,可以将其变为固定形式,如图 1-9 所示。将鼠标移到固定工具栏的拖曳柄处,用鼠标左键向下拖曳或双击,又可将其变为浮动形式。

图 1-8 浮动式窗体编辑器工具栏图　　图 1-9 固定式窗体编辑器工具栏

(2)工具栏的关闭。

① 将工具栏变为浮动形式,单击其右上角的关闭按钮。

② 通过"视图"菜单中的"工具栏"选项或在工具栏上单击鼠标右键,在弹出的快捷菜单中单击带"√"的菜单项,则"√"消失,同时关闭相应工具栏。

4. 工具箱

工具箱用于存放和显示建立 VB 应用程序所需要的各种控件,如图 1-10 所示。

工具箱窗口关闭后再打开的方法有两种。
① 通过"视图"菜单中的"工具箱"选项，如图 1-11 所示。
② 通过"常用工具栏"中的"工具箱"按钮，如图 1-12 所示。

5. 窗体设计器窗口

窗体设计器窗口主要用来设计应用程序的界面。在其工作区内整齐地布满了一些小点，是供设计时对齐控件用的，运行时自动消失。

（1）打开窗体设计器窗口。
① 通过"视图"菜单中的"对象窗口"选项。
② 双击工程资源管理器窗口中相应窗口名称。

图 1-10　工具箱

图 1-11　视图菜单

图 1-12　常用工具栏中的工具箱按钮

（2）将鼠标移到窗体右下角的调整框上，当鼠标指针变为斜向双向键头时，拖曳鼠标调整其宽度为 4000 左右、高度为 3000 左右，如图 1-13 所示。

6. 工程资源管理器窗口

工程资源管理器窗口用于管理工程中所有的文件，以树形目录结构显示，如图 1-14 所示。

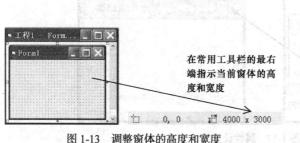

图 1-13　调整窗体的高度和宽度

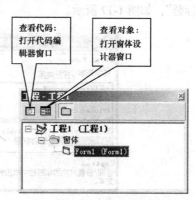

图 1-14　工程资源管理器窗口

（1）打开工程资源管理器窗口，有如下两种方法。

① 通过"视图"菜单中的"工程资源管理器"选项。

② 单击"常用"工具栏中的"工程资源管理器"按钮。

（2）单击"工程"菜单中的"添加窗体"和"添加模块"，向工程 1 中添加另外一窗体 Form2 和一标准模块 Module1，如图 1-15 所示。

（3）在工程资源管理器窗口中双击 Form2 或选中 Form2 后单击上面的"查看对象"按钮，打开 Form2 的窗体设计器窗口。

7. 属性窗口

属性窗口用于设置对象的属性值，如图 1-16 所示。

图 1-15 添加窗体和模块后的工程资源管理器窗口　　　图 1-16 属性窗口

（1）属性窗口的打开，有如下两种方法。

① 通过"视图"菜单中"属性窗口"选项。

② 单击"常用"工具栏中的"属性窗口"按钮。

（2）属性窗口的组成。

① 对象下拉列表：单击其右端向下的箭头可打开下拉列表，列表中列出当前窗体所含对象的名称及类型。

② 选项卡：确定属性的显示方式，即按字母顺序或按分类顺序显示属性。

③ 属性列表框：列出当前对象的所有属性。其中左边为属性名，右边为属性值。

【例题】　将 Form1 的 Caption 属性值设置为"窗体标题"，使窗体运行时其标题显示为"窗体标题"，如图 1-17 所示。

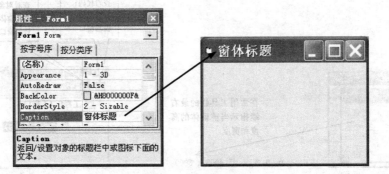

图 1-17 属性设置

8. 窗体布局窗口

窗体布局窗口主要用于调整应用程序窗体运行时在屏幕上的显示位置。

（1）窗体布局窗口的打开，有以下两种方法。

① 通过"视图"菜单中的"窗体布局"选项。

② 单击"常用工具栏"中的"窗体布局"按钮。

（2）用鼠标左键拖动窗体布局窗口中的小窗体图标到屏幕（6045，4110）附近，如图1-18所示。

9. 代码编辑器窗口

代码编辑器窗口是输入应用程序代码的编辑器，如图1-19所示。

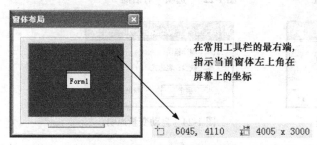
图1-18　窗体在屏幕的坐标

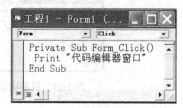
图1-19　代码窗口

代码编辑器窗口的打开有3种方式。

① 通过"视图"菜单中的"代码窗口"选项。

② 双击"窗体设计器"中窗体上的任何位置。

③ 单击"工程资源管理器"中的"查看代码"按钮。

10. 立即窗口、本地窗口和监视窗口

立即窗口、本地窗口和监视窗口是为调试应用程序提供的，如图1-20所示。

图1-20　立即窗口、本地窗口和监视窗口

窗口的打开可分别通过"视图"菜单中的"立即窗口""本地窗口"和"监视窗口"选项。

【例题】　在立即窗口中输入的语句会被立即执行，结果也显示在本窗口中，如图1-21所示。其中Print可以用"?"代替。

11. VB 6.0 集成开发环境的设置

熟悉VB 6.0的集成开发环境后，用户就可以根据自己的喜好和要求配置集成开发环境了。单击"工具"菜单下的"选项"菜单，系统会弹出如图1-22所示的"选项"对话框。

（1）"编辑器"选项卡。

【自动语法检测】　决定输入一行代码后，VB是否应当自动校验语法的正确性。如勾选此选项，如有错，系统将马上显示错误信息，并将有语法错误的语句行用红色显示，如图1-23和图1-24所示。

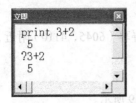 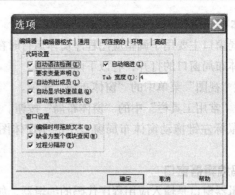

图 1-21 在"立即"窗口中检查表达式的值　　图 1-22 "选项"对话框

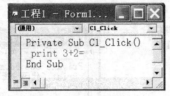 

图 1-23 自动语法检查　　图 1-24 自动变量声明

【自动列出成员】 显示一个框,该框将显示在当前插入点逻辑上完成语句的一些信息。例如,在编写代码时只要输入对象名加".",系统马上列出它具有的成员(如属性和方法等),用户可以从中选取一个成员,如图 1-25 所示。

(2)"通用"选项卡。该选项卡为当前的 Visual Basic 工程指定设置值、错误处理及编译设置值,如图 1-26 所示。

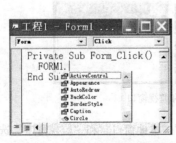 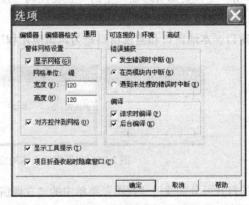

图 1-25 自动列出对象成员　　图 1-26 "选项"对话框中的"通用"选项卡

【显示网格】 决定是否在设计时显示网格。去掉此复选框的选取,则在窗体设计器的窗体上不显示网格点。宽度和高度用于决定窗体上所用网格单元的宽度和高度。

【对齐控件到网格】 如选中此复选框,则向窗体上添加控件时,自动将控件的外部边缘定位在网格线上。

【显示工具提示】 决定是否为工具栏和工具箱各项提供工具提示。

12. 退出 VB

有以下几种方式。

(1)单击窗口右上角的"关闭"按钮,如图 1-27 所示。

（2）选择标题栏左上角的"控制菜单"中的"关闭"选项，如图 1-28 所示。

图 1-27 "关闭"按钮　　图 1-28 "控制菜单"中的"关闭"选项

（3）通过"文件"菜单栏的"退出"选项，如图 1-29 所示。

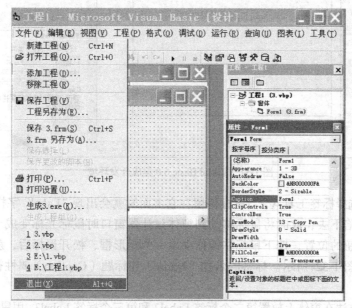

图 1-29 "文件"菜单的"退出"选项

（4）按 Alt+Q 组合键退出。

1.2.2 简单的 Visual Basic 加法程序

【实验要求】

设计一窗体，运行界面如图 1-30 所示。在文本框 Text1 和 Text2 中分别输入"被加数"和"加数"，单击"求和"按钮，求和并显示在 Text3 中；单击"退出"按钮，可结束该程序。

【实验目的】

1. 掌握 VB 6.0 的集成开发环境及各个组成部分的使用方法。

2. 了解程序设计的方法和一般步骤。

【实验步骤】

1. 界面设计

启动 VB 6.0，在"新建工程"对话框中选择"标准 EXE"，单击"打开"按钮，系统将新建

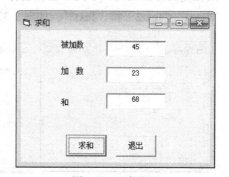

图 1-30 运行界面

一个工程，并进入 VB 6.0 的设计工作模式，此时的默认工程为"工程 1"。该工程只有一个窗体模块，默认窗体为 Form1，如图 1-31 所示。

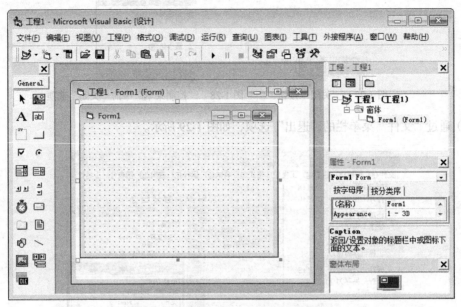

图 1-31　Visual Basic 6.0 集成开发环境

在工具箱中选择标签工具A，当鼠标停留在它上面时就会出现"Label"字样注释。单击它，则图案变亮且凹陷下去，此时将鼠标移动到窗体设计器窗口时鼠标光标变成十字形状。将鼠标移到要放置按钮的位置，按下鼠标左键拖动出一个矩形框，松开鼠标后，就会在窗体上画出一个标签，如图 1-32 所示。标签的名称（Name）和标题（Caption）属性被系统自动设置为"Label1"。

使用同样的方法在窗体上放置第二个标签 Label2 和第三个标签 Label3。用鼠标拖曳框选的办法或按住 Shift 键依次单击，将 3 个标签同时选中，选择"格式"菜单中"统一尺寸"子菜单中的"两者都相同"命令，将标签调成同样大小；选择"格式"菜单中"对齐"子菜单中的"左对齐"命令，将标签左对齐。

在工具箱中选择文本框工具abl，当鼠标停留在它上面时就会出现"TextBox"字样注释。用画标签的方法画出三个文本框，名称（Name）和标题（Caption）属性被系统分别自动设置为"Text1""Text2"和"Text3"，如图 1-32 所示。将三个文本框调成同样大小并左对齐。

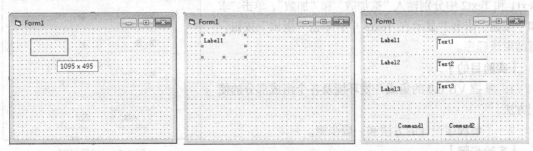

图 1-32　建立标签、文本框和命令按钮

在工具箱中选择命令按钮工具▭，当鼠标停留在它上面时就会出现"CommandButton"字样

注释。用画标签的方法画出两个按钮,名称(Name)和标题(Caption)属性被系统分别自动设置为"Command1"和"Command2",如图 1-32 所示。将按钮调成同样大小,选择"格式"菜单中"对齐"子菜单中的"顶端对齐"命令,将按钮顶端对齐。

2. 属性设置

属性设置如表 2-1 所示。

表 2-1 　　　　　　　　　　　　　　对象属性设置

对象	属性	属性值
Form1	Caption	求和
Label1		被加数
Label2	Caption	加　数
Label3		和
Label1		
Label2	Autosize	True
Label3		
Text1		
Text2	Text	空白
Text3		
Text1		
Text2	Alignment	2-Center
Text3		
Command1	Caption	求和
Command2		退出

(1)在属性窗口中将 Form1 窗体的 Caption 属性设置为"求和",如图 1-33 所示。

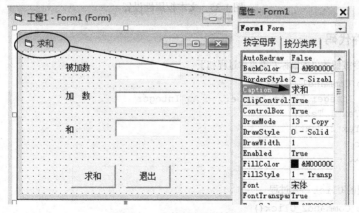

图 1-33　窗体标题属性设置

(2)在属性窗口中将 Label1 标签的 Autosize 属性设置为"True",Caption 属性设置为"被加数",如图 1-34 所示。按同样的方法设置 Label2 和 Label3 的属性。

(3)在属性窗口中将 Text1 文本框的 Text 属性内容清空,Alignment 属性设置为"2-Center",如图 1-35 所示。按同样的方法设置 Text2 和 Text3 的属性。

(4)在属性窗口中将 Command1 按钮和 Command2 按钮的 Caption 属性分别设置为"求和"

和"退出"。

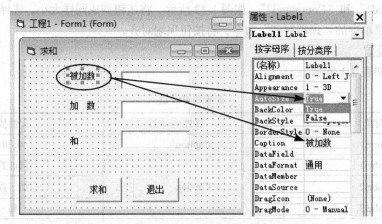

图 1-34　标签属性设置

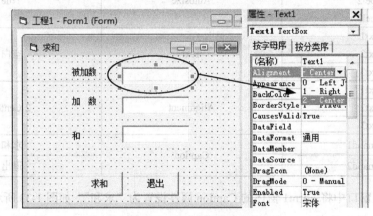

图 1-35　文本框属性设置

3. 编写事件代码

```
' "求和"命令按钮的单击鼠标左键事件过程如下
Private Sub Command1_Click()
    Dim a As Integer, b As Integer, c As Integer
    a = Val(Text1.Text)
    b = Val(Text2.Text)
    c = a + b
    Text3.Text = c
End Sub
' "退出"按钮的 Click 事件过程
Private Sub Command2_Click()
    End
End Sub
```

4. 保存工程，调试程序，生成可执行文件并运行程序

在菜单"文件"中选择"保存工程"选项，或者单击常用工具栏上的保存按钮 🖫。第一次保存文件将弹出"文件另存为"对话框，如图 1-36 所示。在"保存在"下拉列表框中选择文件夹，在"文件名(N)"中输入"1-1.frm"，单击"保存"按钮后，窗体以"1-1.frm"为文件名保存，接着弹出"工程另存为"对话框，如图 1-37 所示。在"保存在"下拉列表框中选择相同的文件夹，

在"文件名(N)"中输入"1-1",工程以"1-1.vbp"文件名保存在磁盘中。

图 1-36 保存窗体文件

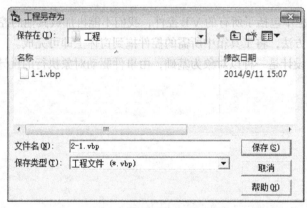

图 1-37 保存工程文件

通过"运行"菜单中的"启动"选项,或单击工具栏上的启动按钮 ▶ 运行程序。

单击"文件"菜单中的"生成 1-1.exe"命令,生成可执行文件。

1.3 测试题

一、选择题

1. 在中断状态下可以运行代码,也可以在运行状态中作为输出结果的窗口称为(   )。
   [A] 本地窗口　　　[B] 立即窗口　　　[C] 观察窗口　　　[D] 栈窗口
2. VB 程序设计采用的编程机制是(   )。
   [A] 可视化　　　　[B] 面向对象　　　[C] 事件驱动　　　[D] 过程结构化
3. 以下叙述中错误的是(   )。
   [A] 打开一个工程文件时,系统自动装入与该工程有关的窗体、标准模块等文件
   [B] 保存 VB 程序时,应分别保存窗体文件及工程文件
   [C] VB 应用程序只能以解释方式执行
   [D] 事件可以由用户引发,也可以由系统引发

4. VB 应用程序保存在磁盘上，至少会有以（   ）为扩展名两个文件。

[A] .DOC 和.TXT   [B] .COM 和.EXE

[C] .VBW 和.BAS   [D] .VBP 和.FRM

5. VB 6.0 是一种面向（   ）的编程环境。

[A] 机器   [B] 对象   [C] 过程   [D] 应用

二、填空题

1. 可以通过_____菜单中的_____命令退出 VB。

2. VB 的工作模式可分为 3 种：设计、运行以及_____。

3. 在 VB 中，要显示程序代码，必须在_____窗口；要设计程序的运行界面，必须在_____窗口。

4. 当进入 VB 集成环境后，发现没有显示"工具箱"窗口，应选择_____菜单的_____选项，使"工具箱"窗口显示。

三、判断题

1. 在 VB 的工具栏中包括了所有的 VB 控件，我们不能再加载其他的控件。   (   )

2. 只要用拖曳的方法，将工具箱中所需的控件拖到窗体上即可完成。   (   )

3. 面向对象程序设计是一种以对象为基础，由事件驱动对象执行的设计方法。   (   )

第 2 章
Visual Basic 应用程序设计过程

2.1 知识点

2.1.1 面向对象程序设计的基本概念

（1）对象：是系统中的基本运行实体。
（2）属性：描述对象特征的数据称为属性。
属性设置有以下两种方法。
① 在属性窗口上设置。
② 在程序代码中用赋值语句设置。
格式：<对象名>.<属性名>=<属性值>
（3）事件过程。
① 事件：系统预先定义好的能够被对象识别的动作。
② 事件过程：对象响应事件后执行的程序代码，这段代码称为事件过程，由用户编写。
事件过程的一般格式（私有）为：
Private Sub <对象名>_<事件名>([<参数列表>])
　　…（事件过程代码）
End Sub
（4）方法：VB 为实现一定的功能而编写的过程。
方法调用的格式为：
　　　[<对象名>.]<方法名>　[<参数>]
（5）类：同种对象的集合，一个对象就是类的一个实例。

2.1.2 窗体

1. 属性

Name：窗体名称。
BackColor：设置窗体的背景颜色。
Caption：设置窗体的标题栏中所显示内容。
Enabled：设置窗体是否有效。为 True 窗体有效，窗体响应用户所产生的事件，为默认值；

为 False 窗体失效，不能响应用户的任何事件或操作。

Font：设置字体、字号等，如表 2-1 所示。

表 2-1　　　　　　　　　　　　　Font 属性设计

FontName	字体（默认为宋体）
FontSize	字体大小
FontBold	是否是粗体
FontItalic	是否是斜体
FontStrikeThru	是否加删除线
FontUnderLine	是否加下划线

Icon：设置窗体左上角的图标。

Picture：设置窗体中要显示的图片。

StartUpPosition：设置窗体启动时所在位置。其值为 0 由窗体的 Left、Top 决定；其值为 1 用户窗体所有者中央；其值为 2 屏幕中央；其值为 3 屏幕的左上角。

Visible 属性：设置窗体运行时是否可见，其值为 True 窗体可见，为默认值；其值为 False 窗体不可见。

WindowState 属性：设置窗体运行时的状态。其值为 0 正常化，为默认值；其值为 1 最小化；其值为 2 最大化。

Left、Top：窗体运行时其左上角距屏幕左端、顶端的距离。窗体在屏幕上的坐标值为（Left, Top）。

Height、Width：窗体的高度和宽度。

2. 事件

Click、DblClick、Load、Unload 等。

3. 方法

Print 方法：在窗体上打印输出数据。

　　　　标准格式：

　　　　　　[对象名.]Print <表达式 1>[, 表达式 2 [, 表达式 3…]]

　　　　紧凑格式：

　　　　　　[对象名.]Print <表达式 1>[; 表达式 2 [; 表达式 3…]]

Cls 方法：清除用 Print 方法在窗体上输出的数据，清除在窗体上绘制的图形。

　　　　格式为：

　　　　　　[对象名.]Cls

Hide 方法：隐藏窗体。格式为：

　　　　　　[窗体名].Hide

Show 方法：显示窗体。格式为：

　　　　　　[窗体名].Show

2.1.3　命令按钮

1. 属性

Style 属性：设置按钮的显示形式。其值 0 为文字按钮，默认值；其值 1 为图形按钮。

Picture：用于添加图形按钮上的图像。

Name、Caption、Font、Enabled 等属性与窗体属性相同。

2. 事件

Click、DblClick、MouseDown、MouseUp 等。

3. 方法

SetFocus：将焦点定位在指定的命令按钮上。

2.1.4 标签

1. 属性

Caption：设置标签所显示内容。

Alignment：对齐方式。设置为 0 为左对齐；为 1 为右对齐；为 2 为居中对齐。

AutoSize：设置是否自动调整标签的大小，其值为 True 可根据标题自动调整标签大小；其值为 False 不自动调整标签大小，为默认值。

BackStyle：用于设置标签是否透明。其值为 1 标签不透明，为默认值；为 0 标签透明。

此外还有 MousePointer、MouseIcon、WordWrap 属性等。

2. 事件

Click、DblClick 等。

3. 方法

Move 等。

2.1.5 文本框

1. 属性

Text：文本框中显示的文本内容。

MultiLine：其值为 False 显示一行文本，为 True 显示多行文本。

MaxLength：文本的最大长度（默认 0，无长度限制）。

PasswordChar：密码设置时，掩盖字符。

Scrollbars：其值为 0 无滚动条；其值为 1 有水平滚动条；其值为 2 有垂直滚动条；其值为 3 有水平及垂直滚动条。

Locked：设置文本是否锁定。其值为 False 没锁定，文本内容可以编辑修改，默认值；为 True 锁定，文本框中文本只能读，不能修改。

Alignment：设置文本对齐方式。其值为 0 文本左对齐；其值为 1 文本右对齐；其值为 2 文本居中对齐。

2. 事件

Change 事件：当文本框内容发生变化时，触发事件。

GotFocus 事件：当文本框得到焦点时，触发该事件。

LostFocus 事件：当文本框失去焦点时，触发该事件。

3. 方法

SetFocus：让文本框获得焦点。

2.1.6 程序设计过程

一个简单程序设计过程的基本步骤如图 2-1 所示。

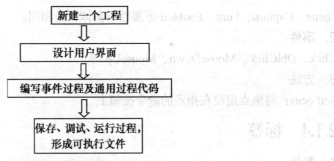

图 2-1　程序设计的基本步骤

2.2　实验内容

2.2.1　单个窗体实验

【实验要求】

设计一个窗体，运行界面如图 2-2 所示。窗体标题设置为"求两个数的和"，其左上角示意图标为 face02.ico（该图标文件一般可以在"Microsoft Visual Studio\ Common\Graphics\Icons\Misc"文件夹下找到，后面各章节实验所需图标均按此路径查找），标签设置背景颜色为绿色及文字的 Font 属性为"黑体三号"字；文本框设置背景颜色为黄色并使文字居中显示；"计算"和"返回"按钮设置图片及访问键，并分别编程实现"求和"功能和"返回"功能；窗体启动时处于屏幕中央。

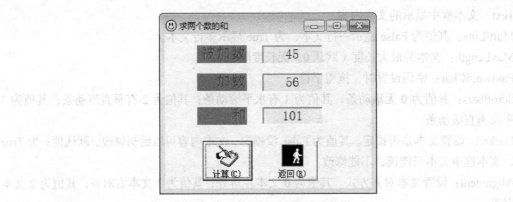

图 2-2　单个窗体实验运行界面

【实验目的】

1. 了解面向对象程序设计的基本原理，理解对象的属性、事件和方法等概念。
2. 掌握窗体及其他控件的常用属性、事件和方法。

【实验步骤】

1. 界面设计

启动 VB 6.0，在"新建工程"对话框中选择"标准 EXE"，单击"打开"按钮，系统将新建一个工程，并进入 VB 6.0 的设计工作模式，此时的默认工程为"工程 1"。该工程只有一个窗体模块，默认窗体为 Form1，如图 2-3 所示。

第 2 章　Visual Basic 应用程序设计过程

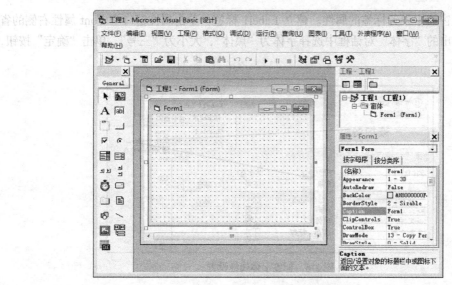

图 2-3　VB 6.0 集成开发环境

2. 属性设置

（1）在属性窗口中将 Form1 窗体的 Caption 属性设置为"求两个数的和"，如图 2-4 所示。

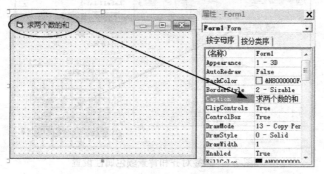

图 2-4　窗体标题属性设置

（2）将 Form1 窗体的 Icon 属性设置为"face02.ico"。单击属性窗口中 Icon 属性右侧的省略号按钮，在弹出的"加载图标"对话框中选择"face02.ico"文件，单击"打开"按钮，将 Form1 窗体的示意图标设置为笑脸图形（该图标文件一般可以在"Microsoft Visual Studio\Common\Graphics\Icons\Misc"文件夹下找到），如图 2-5 所示。

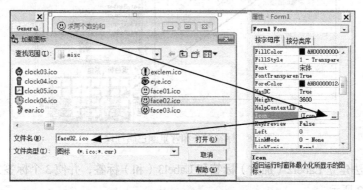

图 2-5　窗体示意图标属性设置

（3）在属性窗口中设置标签的属性。建立 Label1 标签，单击属性窗口中 Font 属性右侧的省略号按钮，在弹出的"字体"对话框中选择字体为"黑体"，大小为"三号"，单击"确定"按钮，如图 2-6 所示。

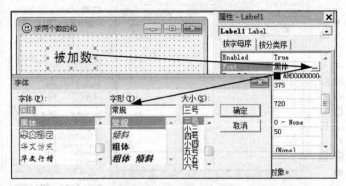

图 2-6　标签字体属性设置

选中 Label1 标签，将属性窗口中 Alignment 属性设置为右对齐"1-Right Justify"，选择 BackColor 属性中的调色板选项卡，将背景色设置为绿色，如图 2-7 所示。

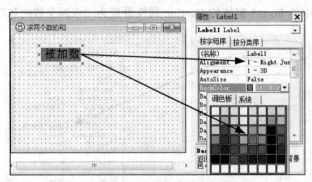

图 2-7　标签对齐和背景颜色属性设置

选中 Label1 标签，选择 ForeColor 属性中的调色板选项卡，将前景色（文字颜色）设置为红色，如图 2-8 所示。

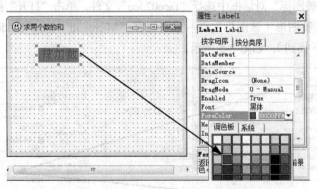

图 2-8　标签对齐和背景颜色属性设置

用同样的方法建立 Label2（加数）标签和 Label3（和）标签，并使 3 个标签大小相同，左对齐，结果如图 2-2 所示。

(4)设置文本框属性。建立 Text1 文本框,用与标签设置相同的方法将其 Font 属性设置为"宋体"、"四号",将其 Alignment 属性设置为居中"2-Center",将其 BackColor 背景色设置为黄色,将其 Text 属性内容清空。

用同样的方法建立 Text2 文本框和 Text3 文本框,并使 3 个文本框大小相同,左对齐。

(5)设置按钮属性。建立 Command1(计算)按钮,将其 Caption 属性设置为"计算(&C)",将其 Style 属性设置为"1-Graphical",单击属性窗口中 Picture 属性右侧的省略号按钮,在弹出的"加载图片"对话框中选择图形"note16.ico"(该图标文件一般可以在"Microsoft Visual Studio\Common\Graphics\Icons\Writing"文件夹下找到),如图 2-9 所示。

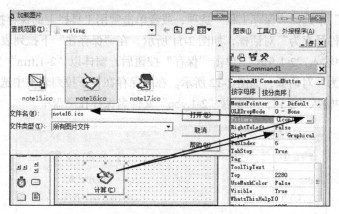

图 2-9 为按钮加载图片

用同样的方法建立 Command2(返回)按钮,单击属性窗口中 Picture 属性右侧的省略号按钮,在弹出的"加载图片"对话框中选择图形"trffc19a.ico"(该图标文件一般可以在"Microsoft Visual Studio\Common\Graphics\Icons\traffic"文件夹下找到)。

(6)将 Form1 窗体启动时在屏幕上显示的位置设置为屏幕中心。在属性窗口中将 StartUpPosition 属性设置为"2-屏幕中心",如图 2-10 所示。

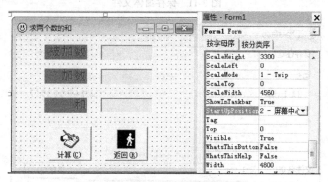

图 2-10 窗体启动时屏幕位置设置

3. 编写事件代码

双击 Command1(计算)按钮进入代码编辑窗口编写程序代码。在 Command1 的 Click 事件中写入如下代码:

```
Private Sub Command1_Click()
    Dim a As Integer, b As Integer, c As Integer
    a = Val(Text1.Text)
```

```
    b = Val(Text2.Text)
    c = a + b
    Text3.Text = c
End Sub
```

双击 Command2（返回）按钮进入代码编辑窗口编写程序代码。在 Command2 的 Click 事件中写入如下代码：

```
Private Sub Command2_Click()
    End
End Sub
```

4. 保存工程

在菜单"文件"中选择"保存工程"选项，或者单击常用工具栏上的保存按钮 。第一次保存文件将弹出"文件另存为"对话框，如图 2-11 所示。在"保存在"下拉列表框中选择文件夹，在"文件名(N)"中输入"2-1.frm"，单击"保存"按钮后，窗体以"2-1.frm"为文件名保存，接着弹出"工程另存为"对话框，如图 2-12 所示。在"保存在"下拉列表框中选择相同的文件夹，在"文件名(N)"中输入"2-1"，工程以"2-1.vbp"文件名保存在磁盘中。

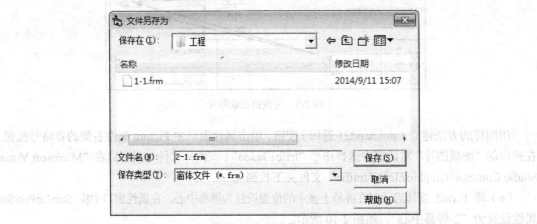

图 2-11 保存窗体文件

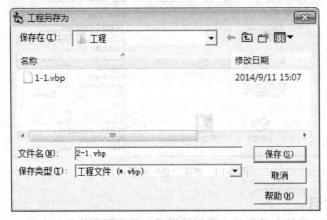

图 2-12 保存工程文件

5. 运行程序、生成可执行文件

通过"运行"菜单中的"启动"选项，或单击工具栏上的启动按钮 ▶ 运行程序。

运行调试程序，直到满意为止。用"文件"菜单生成可执行文件"2-1.exe"。

2.2.2 多个窗体实验

【实验要求】

在实验 2.2.1 设计的工程中增加 4 个窗体，分别完成加、减、乘、除 4 种运算；通过单击"加""减""乘""除"按钮能够显示相应窗体，单击"返回"按钮能返回至四则运算窗体。运行界面如图 2-13 所示。

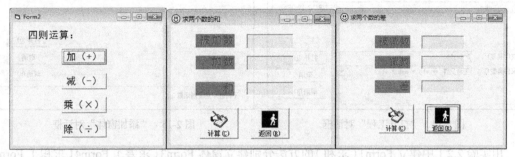

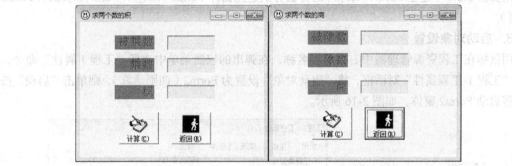

图 2-13 多个窗体实验运行界面

【实验目的】
1. 理解对象的属性、事件和方法等概念。
2. 掌握窗体的添加、显示和隐藏的方法。

【实验步骤】
1. 界面设计

打开实验 2.2.1 所建立的 "2-1.vbp" 工程。

选择"文件"菜单中的"打开工程"选项或单击"常用"工具栏上的打开工程按钮，系统弹出"打开工程"对话框，如图 2-14 所示。然后选择 "2-1.vbp"，单击"打开"按钮，打开实验 2.2.1 所建的工程。

选择"工程"菜单下的"添加窗体"命令，弹出"添加窗体"对话框，如图 2-15 所示，选择"新建"选项卡中的"窗体"选项，单击"打开"按钮，在工程中添加一个窗体 Form2（四则运算）。

使用同样的方法在工程中添加窗体 Form3（求差）、Form4（求积）、Form5（求商）。

2. 属性设置

在 Form2（四则运算）中添加 Label1 标签，在属性窗口中将 Caption 属性设置为"四则运算："，将其 Font 属性设置为"黑体""四号"。

在 Form2（四则运算）中添加 Command1、Command2、Command3、Command4 4 个按钮，在属性窗口中将其 Caption 属性分别设置为"加（+）""减（-）""乘（×）""除（÷）"，将其 Font

属性设置为"黑体""四号"。

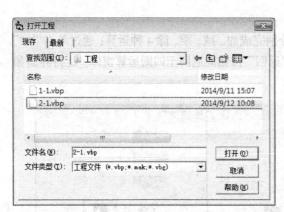

图 2-14 "打开工程"对话框

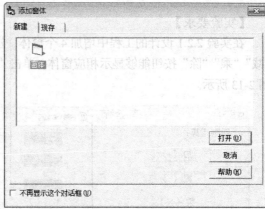

图 2-15 "添加窗体"对话框

用实验 2.2.1 中建立 Form1（求和）的方法分别建立窗体 Form3（求差）、Form4（求积）、Form5（求商）。

3. 启动对象设置

用鼠标在工程资源管理器中右击工程名称，在弹出的快捷菜单中选择"工程 1 属性"命令，弹出"工程 1-工程属性"对话框，将"启动对象"设置为 Form2（四则运算），则单击"启动"按钮时将启动 Form2 窗体，如图 2-16 所示。

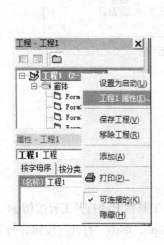
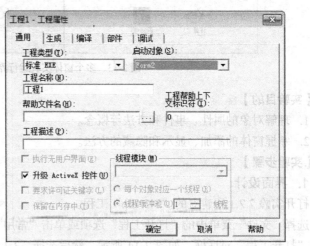

图 2-16 设置启动对象

4. 编写事件代码

（1）Form2 程序清单如下：

```
Private Sub Command1_Click()
    Form1.Show
    Form2.Hide
End Sub

Private Sub Command2_Click()
    Form3.Show
    Form2.Hide
```

```
        End Sub

        Private Sub Command3_Click()
            Form4.Show
            Form2.Hide
        End Sub

        Private Sub Command4_Click()
            Form5.Show
            Form2.Hide
        End Sub
```

（2）Form1 程序清单如下：

```
        Private Sub Command1_Click()
            Dim a As Integer, b As Integer, c As Integer
            a = Val(Text1.Text)
            b = Val(Text2.Text)
            c = a + b
            Text3.Text = c
        End Sub
        Private Sub Command2_Click()
            Form2.Show
            Form1.Hide
        End Sub
```

（3）Form3 程序清单如下：

```
        Private Sub Command1_Click()
            Dim a As Integer, b As Integer, c As Integer
            a = Val(Text1.Text)
            b = Val(Text2.Text)
            c = a - b
            Text3.Text = c
        End Sub
        Private Sub Command2_Click()
            Form2.Show
            Form3.Hide
        End Sub
```

（4）Form4 程序清单如下：

```
        Private Sub Command1_Click()
            Dim a As Integer, b As Integer, c As Integer
            a = Val(Text1.Text)
            b = Val(Text2.Text)
            c = a * b
            Text3.Text = c
        End Sub
        Private Sub Command2_Click()
            Form2.Show
            Form4.Hide
        End Sub
```

（5）Form5 程序清单如下：

```
        Private Sub Command1_Click()
            Dim a As Integer, b As Integer, c As Integer
            a = Val(Text1.Text)
            b = Val(Text2.Text)
            c = a / b
```

```
        Text3.Text = c
End Sub
Private Sub Command2_Click()
    Form2.Show
    Form5.Hide
End Sub
```

5. 保存程序、运行程序

窗体文件分别保存为"2-2-1.frm""2-2-2.frm""2-2-3.frm""2-2-4.frm""2-2-5.frm",工程文件保存为"2-2.vbp"。

运行调试程序,直到满意为止。用"文件"菜单生成可执行文件"2-2.exe"。

2.2.3 文本框实验

【实验要求】

在窗体上放3个文本框Text1、Text2、Text3和2个命令按钮Command1、Command2。运行时,用户在文本框Text1中输入内容的同时,文本框Text2和Text3显示相同的内容,但显示的字体不同,Tex1是宋体、五号字,Text2是楷体_GB2312、四号字;Text3是黑体、三号字;单击"清除"按钮,清空3个文本框中的内容,同时Text1获得焦点;单击"退出"按钮结束程序的运行。运行界面如图2-17所示。

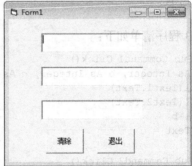

图 2-17　文本框的运行界面

【实验目的】

掌握文本框的 Change 事件。

【实验步骤】

1. 界面设计

新建一个标准 EXE 工程,在窗体上放置 3 个文本框 Text1、Text2、Text3 和两个命令按钮 Command1、Command2,设计成如图 2-17 所示的界面。

2. 属性设置

属性设置如表 2-2 所示。

表 2-2　　　　　　　　　　　　　对象属性设置

对象	属性	属性值
Text1	Text	空白
	Font	宋体、五号字
Text2	Text	空白
	Font	楷体_GB2312、四号字

对象	属性	属性值
Text3	Text	空白
	Font	黑体、三号字
Command1	Caption	清除
Command2	Caption	退出

3. 编写事件代码

```
'编写 Text1 的 Change 事件过程
    Private Sub Text1_Change()
        Text2.Text =Text1.Text
        Text3.Text =Text1.Text
    End Sub
'"清除" 按钮的单击鼠标左键事件过程如下
    Private Sub Command1_Click()
        Text1.Text =""
        Text2.Text =""
        Text3.Text =""
        Text1.SetFocus
    End Sub
'"退出" 按钮的单击鼠标左键事件过程如下
    Private Sub Command2_Click()
        Unload Form1
    End Sub
```

4. 保存工程，调试程序，生成可执行文件并运行程序

以 "2-3.frm" 为窗体文件名，以 "2-3.vbp" 为工程文件名存盘。

运行调试程序，直到满意为止。生成 "2-3.exe" 可执行文件。

2.3 测试题

一、选择题

1. 当事件能被触发时，（　　）就会对该事件作出响应。

   [A] 对象　　　　[B] 程序　　　　[C] 控件　　　　[D] 窗体

2. 下列说法正确的是（　　）。

   [A] 对象属性只能在 "属性窗口" 中设置

   [B] 一个新的工程可以在 "工程窗口" 中建立

   [C] 必须先建立一个工程，才能开始设计应用程序

   [D] 只能在 "代码窗口" 中编写程序代码

3. VB 6.0 中任何控件都有的属性是（　　）。

   [A] BackColor　　[B] Caption　　[C] Name　　[D] BorderStyle

4. 文本框的返回值的类型为（　　）。

   [A] 数值型　　　[B] 变体类型　　[C] 字符串型　　[D] 日期型

5. 在 VB 中最基本的对象是（　　），它是应用程序的基石，是其他控件的容器。
   　　[A] 文本框　　　　[B] 命令按钮　　　　[C] 窗体　　　　[D] 标签
6. 下面（　　）控件不具有 Caption 属性。
   　　[A] 标签框　　　　[B] 单选钮　　　　[C] 命令按钮　　　　[D] 文本框
7. 要使标签能透出窗体的背景，必须设置（　　）属性。
   　　[A] BackStyle　　　[B] BorderStyle　　　[C] Appearance　　　[D] BackColor
8. 要使得标签能自动扩充以满足字体大小则可对其（　　）属性进行设置。
   　　[A] alignment　　　[B] usemnemonic　　　[C] autosize　　　[D] tag
9. 若要使命令按钮不可操作，要对（　　）属性进行设置。
   　　[A] Enabled　　　[B] Visible　　　[C] BackColor　　　[D] Caption
10. 要使得窗体在出现之前就完成相关的程序设置可在（　　）事件中进行编程。
    　　[A] linkopen　　　[B] KeyPress　　　[C] load　　　[D] click
11. 程序运行后，在窗体上单击鼠标，此时窗体不会接收到的事件是（　　）。
    　　[A] MouseDown　　[B] MouseUp　　[C] Load　　[D] Click
12. 要使文本框获得输入焦点，则应采用文本框控件的哪个方法（　　）。
    　　[A] GodFocus　　[B] LostFocus　　[C] KeyPress　　[D] SetFocus
13. 能够获得一个文本框中被选取文本的长度的属性是（　　）。
    　　[A] Text　　　　[B] SelLength　　　[C] SelStart　　　[D] Seltext

二、填空题
1. 在 VB 中，事件过程名由_____和事件构成。
2. 在 VB 中，若要改变一个窗体的标题，应在属性窗口中改变这个窗体的_____属性的值。
3. 一个控件在窗体上的位置由 Top 和_____属性决定，其大小由 Width 和_____属性决定。
4. 在 VB 中设置或修改一个对象的属性的方法有两种，它们分别是属性窗口设置和_____。
5. 要使一个按钮设为默认按钮，应把此按钮的_____属性的值设为 True。
6. 在 VB 中，若要使一个文本框（Text）中的内容在超过文本框的宽度时能够自动换行显示，应当将这个文本框的_____属性的值设置为 True。
7. 当文本框的值发生改变时触发事件_____。
8. 文本框的 Locked 属性用来决定文本内容是否能被编辑，当其值为_____时，可以编辑，当其值为_____时，不可以编辑。
9. 设置控件背景颜色的属性名称是_____。

三、判断题
1. 程序启动后标签控件可以用来让用户输入数据。　　　　　　　　　　　　（　　）
2. 每个对象都有一系列预先定义好的对象事件，但要使对象能响应具体的事件，则编写该对象相应的事件过程。　　　　　　　　　　　　　　　　　　　　　　　　（　　）
3. 一个程序只能有一个窗体。　　　　　　　　　　　　　　　　　　　　　（　　）
4. 对象的三要素包括：属性、事件、方法。　　　　　　　　　　　　　　　（　　）
5. VB 若同一个工程有许多窗体，可以指定任一窗体为启始窗体。　　　　　（　　）

28

第 3 章
VB 语言基本知识

3.1 知识点

3.1.1 Visual Basic 的数据类型

VB 的数据类型如表 3-1 所示。

表 3-1　　　　　　　　　　　　　VB 的数据类型

数据类型		关键字	声明变量类型符	变量的初值
数值型	整型	Integer	%	0
	长整型	Long	&	0
	单精度	Single	!	0
	双精度	Double	#	0
	货币型	Currency	@	0
	字节型	Byte		0
字符型		String	$	""
日期型		Date		#0:00:00#
逻辑型		Boolean		False
变体型		Variant		""
对象型		Object		

3.1.2 常量和变量

（1）常量：具体的量，如 12，"你好"，True，#2010-8-28#。
（2）定义变量：Dim| Private| Public| Static <变量名> [As <数据类型>]。
（3）变量的隐式声明：事先不定义数据类型，使用时根据对变量的赋值来决定数据类型。
（4）用符号声明变量：如表 3-1 所示。

3.1.3 运算符与表达式

（1）算术运算：（　）、^、*、/、\（整除）、Mod（求余）、+、-。

（2）字符串运算：&、+。

（3）关系运算：<、<=、>、>=、=、<>。

日期的大小：未来的日期比过去的日期大。

字符按 ASCII 值比大小，从左开始逐个比较。

字符 ASCII 值："　"<"0"<"1"<……<"9" <"A"<"B"<……"Z"<"a"<"b"<……<"z"。

（4）布尔运算符：NOT、AND、OR。

3.1.4　常用内部函数

1. 数学函数

（1）Fix(x)　　截取参数的整数部分函数
　　　例如：　Fix(-2.89)　　结果为-2

（2）Int(x)取整(取小于或等于参数的最大整数)函数
　　　例如：　Int(45.67)　　结果为 45
　　　　　　　Int(-2.89)　　结果为-3

（3）Abs(x)　　绝对值函数
　　　例如：　Abs(-4)　　结果为 4

（4）Rnd　随机产生 0～1 的单精度值
　　　例如：　Int(6*Rnd)+1　结果为 1～6 之间的任意整数。
　　生成[a,b] 范围内的随机整数：Int((b-a+1)*Rnd+a)。
　　初始化随机数发生器：Randomize。

（5）Round(x,n)　　按小数位数四舍五入
　　　例如：　Round(3.14159,3)　结果为 3.142

（6）Sqr(x)　　算术平方根
　　　例如：　Sqr(9)　　结果为　3

（7）Sgn(x)　　取参数的符号值（参数大于 0，返回 1，小于 0，返回-1，等于 0，返回 0）
　　　例如：　Sgn(8.8) 结果为　1

（8）Exp(x)　　自然指数
　　　例如：　Exp(2)　　等于　　　e*e

（9）Log(x)　　常用对数（求自然对数值）
　　　例如：　Log(1)　结果为　　0

（10）Sin(x)　　正弦；Cos(x) 余弦；Tan(x)　正切；Atn(x)　　计算反正切
　　　例如：　Tan(3.14159265/180*45)　　结果为　1（45 度的正切值）

2. 字符串函数

（1）Len(x)　　求字符串的长度(字符个数)
　　　例如：　Len("ABCDE")　　结果为 5

（2）Mid(x,n1,n2)　从 x 字符串左边第 n1 个位置开始向右取 n2 个字符
　　　例如：　Mid("ABCDE",2,3)　　结果为"BCD"

（3）Instr(x1,x2,M)　返回字符串 x2 在字符串 x1 中的位置，找不到则返回 0，M=1 不区分大小写,省略则区分。
　　　例如：　InStr("ABCDE","D")　　结果为 4

（4）Ltrim(x)　　去掉 x 字符串左边的空格

　　例如：　　LTrim("∪∪∪ABC")　　结果为"ABC"

（5）Rtrim(x)　　去掉 x 字符串右边的空格

　　例如：　　RTrim("ABC∪∪")　　结果为"ABC"

　　例如：　　Trim("∪∪ABC∪∪")　　结果为"ABC"

（6）Left(x,n)　　从 x 字符串左边开始取 n 个字符

　　例如：　　Left("ABCDE",2)　　结果为"AB"

（7）Right(X,n) 从 x 字符串右边开始取 n 个字符

　　例如：　　Right("ABCDE",2)　　结果为"DE"

（8）UCase(x)　　将 x 字符串中所有小写字母转换为大写

　　例如：　　UCase("Abc")　　结果为"ABC"

（9）LCase(x)　　将 x 字符串中所有大写字母转换为小写

　　例如：　　LCase("ABc")　　结果为"abc"

（10）String(n,x)　　返回由 n 个首字符组成的字符串

　　例如：　　String(3, "ABC")　　结果为"AAA"

（11）Space(n) 返回 n 个空格

　　例如：　　Spc(n)：插入 n 个空格　　Tab(n)：定位在 n 列。

3. 日期和时间函数

（1）Date 返回系统日期

　　例如：　　Date()　　返回系统日期格式为(yyyy-mm-dd)

（2）Time 返回系统时间

　　例如：　　Time()　　返回系统时间格式为(hh:mm:ss)

（3）Now 返回系统日期和时间

　　例如：　　Now 返回系统日期时间格式为(yyyy-mm-dd hh:mm:ss)

4. 转换函数

（1）　Str(x)　　将数值转换为字符串

　　例如：　　Str(45.2)　　结果为　"45.2"

（2）Val(x)　　将字符串中的数字转换成数值

　　例如：　　Val("2.3ab")　　结果为　2.3

　　　　　　　Val("a23")　　结果为 0

（3）Asc(x)　　求字符 ASCII 值

　　例如：　　Asc("a")　　结果为 97

（4）Chr(x)　　将数值（ASCII 码）转换为字符

　　例如：　　Chr(65)　　结果为　　　"A"

（5）CBool(x)　　将数字字符串或数值转换成布尔型

　　例如：　　CBool(1)　　结果为 True

　　　　　　　CBool("0")　　结果为 False

　　注：等于 0 为 False，不等于 0 为 True

（6）CDate(x)　　将有效的日期字符串转换成日期

　　例如：　　CDate(#1990,2,23#)　　结果为"1990-2-23"

（7）CSng(x)　　将数值转换成单精度型
　　　例如：　CSng(23.5125468)　　　结果为 23.51255
（8）CDbl(x)　　将数值转换成双精度型
　　　例如：　CDbl(23.5125468)　　　结果为 23.5125468

3.2　实验内容

Visual Basic 的变量、表达式、函数实验

【实验目的】
1. 了解 VB 基本数据类型的定义符、定义关键字和存储数值范围。
2. 掌握变量的声明、赋值和输出方法。
3. 掌握算术表达式、关系表达式和逻辑表达式的正确书写格式。
4. 掌握常用标准数学函数和字符串函数的用法。

一、程序改错

第一题

【实验要求】

设计如图 3-1 所示的窗体，编写家庭房屋装修中用于计算墙面瓷砖数量的小程序。程序运行时，瓷砖的长度和宽度必须填写，每片瓷砖单价以及墙面的高度、长度和宽度可以选择填写，不填写时默认为 0，单击"计算"按钮，计算瓷砖横贴和竖贴时购买瓷砖的精确总数和总价。单击"清空"按钮，所有文本框内容清空。根据程序功能要求修改所给程序段中的错误并运行程序。

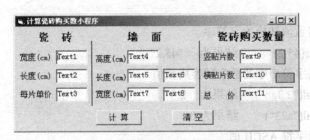

图 3-1　计算瓷砖数程序窗体设计界面

数据输入功能要求：Text1 中瓷砖宽度输入结束后按回车键，光标自动跳转到 Text2 中继续输入，依次类推，直至 Text8 中按回车键后，光标跳转到"计算"按钮。

【实验步骤】

1. 界面设计

在窗体上添加 11 个标签 Label1…Label11，添加 11 个文本框 Text1…Text11，添加 2 个命令按钮 Command1 和 Command2。添加瓷砖横贴和竖贴图形标记及 3 栏间的分割线。

2. 属性设置

对象属性如表 3-2 所示。

表 3-2　　　　　　　　　　　　　　对象属性设置

对　　象	属　性	属　性　值
Command1	Caption	计算
Command2	Caption	清空
Text1～Text11	Text	空
	Alignment	1-Right Justfy
Text1～Text8 Command1、Command2	TabIndex	1、2、3…8 9　10

3. 编写事件代码

```
Private Sub Command1_Click()
    Dim a1%, a2%, a3%, a4%, b1%    'b1存储墙高度，a1~a4存储墙长度和宽度
    Dim Lenth%, width%             ' Lenth 和 width存瓷砖长和宽
    Dim num1%, num2%      '横贴和竖贴时瓷砖数量
    Dim price@        '瓷砖每片单价
    width = Val(Text1.Text)
    Lenth = Val(Text2.Text)
    If (Lenth <> 0 And width <> 0) Then
        b1 = Val(Text4.Text)
        a1 = Val(Text5.Text)
        a2 = Val(Text6.Text)
        a3 = Val(Text7.Text)
        a4 = Val(Text8.Text)
'提醒：贴瓷砖时一面墙横向贴5块瓷砖时，如果还剩余3厘米以上，
'一行要买6块瓷砖，动脑筋学会用整除运算符
        '**********FOUND**********
        num1 = (b1 / width) * _
        (a1 / Lenth + a2 / Lenth + a3 / Lenth + a4 / Lenth)
        '**********FOUND**********
        num2 = (b1 / Lenth) * _
        (a1 / width + a2 / width + a3 / width + a4 / width)
        Text9 = num1
        Text10 = num2
        price = Val(Text3.Text)
        '**********FOUND**********
        Text11 = "横贴: " & num1 * price & "或竖贴: " & num2 * price
    End If
End Sub
Private Sub Command2_Click()
    Text1.Text=""
    Text2.Text=""
    '下面相似语句补充完整
'********** Program **********

'********** End **********
End Sub

'Text1中输入数值按回车键后，光标跳到Text2中
Private Sub Text1_KeyPress(KeyAscii As Integer)
    If KeyAscii = 13 Then   '若所按键的ASCII码值是13，即按了回车键
```

```
        Text2.SelStart = 0
        Text2.SelLength = Len(Text2.Text)   'Text2 中文字被选中,便于修改
        Text2.SetFocus   'Text2 获取焦点,可以优先输入文字
    End If
End Sub
'Text2-text8 的输入功能与 text1 功能相同,程序段相似,同学自行添加。
```

第二题

【实验要求】

设计如图 3-2（a）所示的窗体,在文本框中输入二元一次方程系数,单击"求方程的根"按钮,计算 b^2-4ac 的值,大于等于 0 时计算并输出两个实根,如图 3-2（b）所示,小于 0 时输出方程无实根。根据程序功能修改所给程序段中的错误并运行程序。

   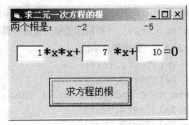

（a）程序设计界面　　　　　　　　（b）程序运行结果界面

图 3-2　求方程的根界面图

【实验步骤】

1. 界面设计

在窗体上添加三个标签 Label1、Label2、Label3,添加 3 个文本框,控件名从左到右分别是 Text1、Text2、Text3,添加 1 个命令按钮 Command1。

2. 属性设置

对象属性设置如表 3-3 所示。

表 3-3　　　　　　　　　　　　对象属性设置

对象	属性	属性值
Text1、Text2、Text3	Text	空
	Alignment	1 Right Juotfy
Label1、Label2、Label3	Caption	*x*x+、*x+、=0
Command1	Caption	求方程的根
Text1、Text2、Text3、Command1	TabIndex	1、2、3、4

3. 编写事件代码

```
Private Sub Command1_Click()
    Dim a%, b%, c%, dt%, x1!, x2!
    a = Val(Text1.Text)
    b = Val(Text2.Text)
    c = Val(Text3.Text)
'**********FOUND**********
    dt =b*b-4ac
```

```
    If (dt > 0) Then
    '**********FOUND**********
      x1 =-b+sqrt(dt)/2*a
    '**********FOUND**********
      x2 =-b-sqrt(dt)/2*a
      Print "两个根是: ", x1, x2
    Else
      Print "方程无根"
    End If
End Sub
```

二、程序填空

第一题

【实验要求】

设计如图3-3所示的窗体，程序运行时在文本框text1中输入一个整数，或单击"产生随机4位整数"获取一个随机4位整数，单击"数值正负取反"按钮是将text1中数值取反，单击"结果保留2位小数"按钮是按四舍五入原则将Text2中数值保留2位小数，单击其他3个按钮，用函数实现各命令按钮的相应功能，结果存放在text2中。将所给程序段的【?】处填写完整并运行程序。

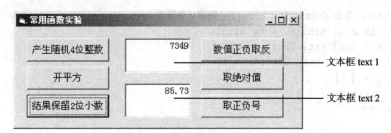

图3-3 函数实验运行结果界面

【实验步骤】

1. 界面设计

在窗体上添加2个文本框Text1、Text2；6个命令按钮Command1～Command6。其大小及位置关系如图3-3所示。

2. 属性设置

对象属性设置如表3-4所示。

表3-4　　　　　　　　　　　　对象属性设置

对象	属性	属性值
Text1、Text2	Text	空
	Alignment	1-Right Justfy
Command1 Command2 Command3 Command4 Command5 Command6	Caption	产生随机4位整数 开平方 结果保留2位小数 数值正负取反 取绝对值 取正负号

3. 编写事件代码

```
Private Sub Command1_Click()    '产生随机4位整数
    Text1.Text = Int(Rnd() * 9000 + 1000)
End Sub

Private Sub Command2_Click()    '开平方
    Dim x As Integer, y As Single
    x = Val(Text1.Text)
    '**********SPACE**********
    y = 【?】
    Text2.Text = y
End Sub

Private Sub Command3_Click()    '结果保留2位小数
    Dim x As Single, y As Single
    x = Val(Text2.Text)
    '**********SPACE**********
    y = 【?】
    Text2.Text = y
End Sub

Private Sub Command4_Click()    '数值正负取反
    Dim x As Single, y As Single
    x = Val(Text1.Text)
    '**********SPACE**********
    y = 【?】
    Text1.Text = y
End Sub

Private Sub Command5_Click()    '取绝对值
    Dim x As Single, y As Single
    x = Val(Text1.Text)
    '**********SPACE**********
    y = 【?】
    Text2.Text = y
End Sub

Private Sub Command6_Click()    '取正负号
    Dim x As Single, y As Single
    x = Val(Text1.Text)
    '**********SPACE**********
    y = 【?】
    Text2.Text = y
End Sub
```

第二题

【实验要求】

设计如图 3-4（a）所示的窗体，练习使用字符串函数实现每个命令按钮要求的功能，单击"查找字符串"按钮时，弹出如图 3-4（b）所示窗口，输入查找字符，将所给程序段的【?】处填写完整并运行程序。

第 3 章 VB 语言基本知识

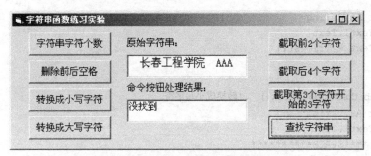

（a）字符串函数练习运行结果界面

（b）查找字符输入界面

图 3-4 字符串函数练习界面

【实验步骤】
1. 界面设计

在窗体上添加 2 个标签，标签名是 Label1、Label2；添加 2 个文本框 Text1 和 Text2，添加 8 个命令按钮 Command1、Command2…Command8，其大小及位置关系如图 3-4（a）所示。各对象属性设置如表 3-5 所示。

表 3-5　　　　　　　　　　　　　对象属性设置

对象	属性	属性值
Text1、Text2	Text	空
	Alignment	1-Right Justfy
Command1	Caption	字符串字符个数
Command2		删除前后空格
Command3		转换成小写字符
Command4		转换成大写字符
Command5		截取前 2 个字符
Command6		截取后 4 个字符
Command7		截取第 3 个字符开始的 3 字符
Command8		查找字符串

2. 编写事件代码

```
Private Sub Command1_Click()    '字符串字符个数
    Dim str As String
    str = Text1.Text
    Text2.Text = Len(str)
End Sub


Private Sub Command2_Click()    '删除前后空格
```

```
        Dim str As String
        str = Text1.Text
        Text2.Text = Trim(str)
    End Sub

    Private Sub Command3_Click()      '转换成小写字符
        Dim str As String
        str = Text1.Text
        '**********SPACE**********
        Text2.Text = 【?】
    End Sub

    Private Sub Command4_Click()      '转换成大写字符
        Dim str As String
        str = Text1.Text
        '**********SPACE**********
        Text2.Text = 【?】
    End Sub

    Private Sub Command5_Click()      '截取前2个字符
        Dim str As String
        str = Text1.Text
        '**********SPACE**********
        Text2.Text = 【?】
    End Sub

    Private Sub Command6_Click()      '截取后4个字符
        Dim str As String
        str = Text1.Text
        '**********SPACE**********
        Text2.Text = 【?】
    End Sub

    Private Sub Command7_Click()      '截取第3个字符开始的3字符
        Dim str As String
        str = Text1.Text
        '**********SPACE**********
        Text2.Text = 【?】
    End Sub

    Private Sub Command8_Click()      '查找字符串
        Dim str As String
        str = InputBox("请输入要查找的字符串", "查找")
        '**********SPACE**********
        Text2.Text = 【?】
        If (Text2.Text = "0") Then
            Text2.Text = "没找到"
        End If
    End Sub

    Private Sub Form_Load()
```

```
    Text1.Text = "  长春工程学院   AAA "
End Sub
```

三、编程题

【实验要求】

设计如图 3-5（a）所示的教师基本信息输入窗口，单击"核对"按钮，弹出 form2 窗体，单击"打印预览"按钮，将核对信息模拟打印在屏幕上，如图 3-5（b）所示。单击"返回"按钮，form2 隐藏，主窗体显示；单击"保存"按钮，定义变量，然后将文本框中的教师基本信息暂时保存到变量中（学习文件一章后，将变量中数据保存在文件中）；单击"清空"按钮，窗口文本框信息清空。根据程序功能编写程序。

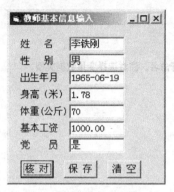

（a）程序运行初始界面　　　　　（b）程序运行结果界面

图 3-5　教师基本信息界面

【实验步骤】

1. 界面设计

在 form1 窗体上添加 7 个标签，标签名从上到下依次是 Label1、Label2…Label7；添加 7 个文本框，文本框名从上到下依次是 Text1、Text2…Text7，添加 3 个命令按钮 Command1、Command2 和 Command3。

在工程中添加 form2 窗体，在 form2 窗体中添加 2 个命令按钮 Command1 和 Command2。

2. 主要属性设置

属性设置如表 3-6、表 3-7 所示。

表 3-6　　　　　　　　　　　　　form1 对象属性设置

对象	属性	属性值
Text1、Text3、Text4、Text5、Text6	Text	空
Text1、Text2～Text7、Command1、Command2、Command3	TabIndex	1、2…6、7 8、9、10
Text2	Text	男
Text7	Text	是
Command1	Caption	核对
Command2	Caption	保存
Command3	Caption	清空

表 3-7　　　　　　　　　　　　form2 对象属性设置

对象	属性	属性值
Command1、Command2	TabIndex	1、2
Command1	Caption	打印预览
Command2	Caption	返回

3. 编写事件代码

```
'Form1 窗体代码
Option Explicit
Private Sub Command1_Click()
   Form2.Show
End Sub
Private Sub Command2_Click()
     '声明变量,变量名用英文单词或汉语拼音头写字符命名,注意选择变量数据类型。
   Dim name As String
   Dim Sex As String
   '********** Program *********



'********** End ************
   '教师信息暂存变量中
   name = Text1.Text
   Sex = Text2.Text
   '********** Program *********



'********** End ************
End Sub
Private Sub Command3_Click()
   Text1.Text = "":       Text2.Text = "男":      Text3.Text = ""
   Text4.Text = "":       Text5.Text = "":        Text6.Text = ""
   Text7.Text = "否"
End Sub
'Form2 窗体代码
Private Sub Command1_Click()    '打印预览按钮
    Print Tab(3); "请核对以下信息："
    Print Tab(3); "姓名："; Form1.Text1.Text
    Print Tab(3); "性别："; Form1.Text2.Text
    '********** Program *********



'********** End ************
    Print Tab(3); Date
End Sub
Private Sub Command2_Click()    '返回按钮
    '********** Program *********
```

```
'********** End **************
End Sub
```

3.3 测 试 题

一、选择题

1. 在 VB 中，下列两个变量名相同的是（　　）。
   [A] Knol 和 Kn_ol [B] CHINA 和 china
   [C] English 和 Engl.sh [D] China 和 Chin
2. 下面（　　）是合法的字符常数。
   [A] ABC$ [B] "ABC" [C] 'ABC' [D] ABC
3. 下列哪个变量名的命名是不正确的（　　）。
   [A] TName [B] T_Temp [C] T32 [D] 23T#
4. 如果一个整型变量定义后未赋值，则该变量的缺省值为（　　）。
   [A] 8 [B] "" [C] 1 [D] 0
5. 可以在变量的后面加上类型说明符以显示变量的类型，可以表示整型变量的是（　　）。
   [A] % [B] # [C] ! [D] $
6. 要定义一个变量为全局变量，应使用（　　）关键字。
   [A] Static [B] Public [C] Private [D] Sub
7. 下列对变量的定义中，不能定义 A 为变体变量的是（　　）。
   [A] DIM A AS DOUBLE [B] DIM A AS VARIANT
   [C] DIM A [D] A=24
8. 如果一个变量未经定义而直接使用，则该变量的类型为（　　）。
   [A] Integer [B] Byte [C] Boolean [D] Variant
9. OPTION　EXPLICIT 语句不可以放在（　　）。
   [A] 窗体模块的声明段中 [B] 标准模块的声明段中
   [C] 类模块的声明段中 [D] 任何事件过程中
10. 表达式"456"& 123 的结果是（　　）。
    [A] "456123" [B] 579 [C] 语法错误 [D] 456123
11. 产生[10～37]的随机整数的 VB 表达式是（　　）。
    [A] Int(Rnd(1)*27)+10 [B] Int(Rnd(1)*28)+10
    [C] Int(Rnd(1)*27)+11 [D] Int(Rnd(1)*28)+11
12. 表达式"xyz" + "568"的值是（　　）。
    [A] xyz [B] 568 [C] xyz568 [D] 120
13. 表达式(9\3+1)*(15\2+2)的值为（　　）。
    [A] 36 [B] 38 [C] 21.4 [D] 8.4
14. 代数式 e*Sin(30°)2x/(x+y)lnx 对应的 VB 表达式是（　　）。

[A] E^X*Sin(30*3.14/180)*2*x/x+y*log(x)

[B] Exp(x)*Sin(30)*2*x/(x+y)*ln(x)

[C] Exp(X)*Sin(30*3.14/180)*2*x/(x+y)*log(x)

[D] Exp(X)*Sin(30*3.14/180)*2*x/(x+y)*ln(x)

15. 已知 A、B、C 中 C 最小，则判断 A、B、C 可否构成三角形三条边长的逻辑表达式是（　　）。

    [A] A>=B And B>=C And C>0

    [B] A+C>B And B+C>A And C>0

    [C] (A+C)>=C And A-C <=C) And C>0

    [D] A+B>C And A-B>C And C>0

16. 表达式 INT(8*SQR(36)*10^(-2)*10+0.5)/10 的值是（　　）。

    [A] 0.48　　　　[B] 0.048　　　　[C] .5　　　　[D] .05

17. 下列表达式（　　）是不正确的。

    [A] "计算机"+"与程序设计"　　　　[B] 计算机+150

    [C] "计算机"&150　　　　[D] "计算机"&"与程序设计"

18. 表达式 val(a2000)的值为（　　）。

    [A] 0　　　　[B] 65　　　　[C] 2000　　　　[D] 19

19. 表达式(-1)*Sgn(-100+Int(Rnd*100))的值是（　　）。

    [A] 0　　　　[B] 1　　　　[C] -1　　　　[D] 随机数

20. Int(198.555*100+0.5)/100 的值为（　　）。

    [A] 198　　　　[B] 199.6　　　　[C] 198.56　　　　[D] 200

21. 下面选项中（　　）是字符连接运算符。

    [A] %　　　　[B] Mod　　　　[C] &　　　　[D] And

22. \、/、Mod、*4 个算术运算符中，优先级最低的是（　　）。

    [A] \　　　　[B] /　　　　[C] Mod　　　　[D] *

23. 下列语句的输出结果是（　　）。

Print Format (1234.56, "##,###.#")

    [A] 1,234. 6　　　[B] 01,234.56　　　[C] 1234.6　　　[D] 1,234.56

24. 下列语句的输出结果为（　　）。

Print Format$(5689.36, "00,000.00")

    [A] 5,689.36　　　[B] 5,689.360　　　[C] 05,689.36　　　[D] 005,689.360

25. 语句 PRINT"INT(-13.2)= ";INT(-13.2)的输出结果为（　　）。

    [A] INT(-13.2)= -13.2　　　　[B] INT(-13.2)=13.2

    [C] INT(-13.2)= -13　　　　[D] INT(-13.2)=-14

26. 假设 A="How are you"，B="a Boy and a girl"，则下列表达式的结果等于"are you a girl"的是（　　）。

    [A] Right(A,7)+Mid(B,10,7)　　　　[B] Left (A,7)+Mid(B,10,6)

    [C] Mid(B,10,6)+Right(A,7)　　　　[D] Right(A,7)+Mid(B,10,6)

27. 假设 A="Good Morning."，B="Afternoon, Boys."，则下列表达式的结果等于"Good Boys."的是（　　）。

[A] Left(A,5)+Right(B,5)   [B] Left(A,10)+Right(B,6)
[C] Mid(A,1,5)+Mid(B,1,6)   [D] Mid(A+B,1,11)

28. 函数 Len(Str(Val("123.4"))) 的值为（　　）。
   [A] 11   [B] 5   [C] 6   [D] 8

29. 下列字符串常量中，最大的是（　　）。
   [A] "北京"   [B] "上海"   [C] "天津"   [D] "广州"

30. 设 a=10，b=5，c=1，执行语句 Print a>b>c 后，窗体上显示的是（　　）。
   [A] True   [B] False   [C] 1   [D] 出错信息

31. 如果 TAB 函数的参数小于 1，则打印位置在第（　　）列。
   [A] 0   [B] 1   [C] 2   [D] 3

32. 以下几项中，属于逻辑型常量的是（　　）。
   [A] Bal   [B] 2010-10-2   [C] #10/10/02#   [D] TRUE

33. 下列几项中，属于合法的日期型常量的是（　　）。
   [A] "23/12/02"   [B] 23/12/02   [C] {23/12/02}   [D] #23/12/02#

34. 下列语句中，不能交换变量 a 和 b 的值的是（　　）。
   [A] t=b : b=a : a=t   [B] a=a+b : b=a-b : a=a-b
   [C] t=a : a=b : b=t   [D] a=b : b=a

二、填空题

1. 声明定长为 10 个字符变量 Sstr 的语句为_____。

2. 表达式 1 and 0 的值为_____。

3. Integer 类型的变量占用_____字节空间。

4. 已知 a=3.5，b=5.0，c=2.5，d=True，则表达式：a>=0 AND a+c>b+3 OR NOT d 的值是_____。

5. 设 x=2，y=4，a=4，b=6，表达式 (a>x Xor b<x) OR x<y IMP y<b 的结果为_____。

6. 在 VB 中，若要将字符串"12345"转换成数字值应使用的类型转换函数是_____。

7. 一元二次方程 $ax^2+bx+c=0$ 有实根的条件为 a≠0，并且 $b^2-4ac≥0$，列出逻辑表达式_____。

8. X=2:Y=8:PRINT X+Y=10 的结果是_____。

9. 变量 min@ 表示_____类型的变量。

10. VB 中，变量的声明可分为两种方法：隐式声明和_____。

11. 要从字符串"Made In China"中截取子字符串"In"可使用函数_____。

12. 表达式 Ucase(Mid("abcdefgh",3,4)) 的值是_____。

13. 如果要指明 mystring 为固定 15 个字长的全局变量，应该在标准模块中用以下说明语句定义_____。

14. 声明单精度常量 P1 代表 3.14159 的语句为_____。

15. 把整数 0 赋给一个逻辑型变量，则逻辑变量的值为_____。

16. VB 表达式 9^2 MOD 45 \2 *3 的值为_____。

17. 变量 L 的值为−8，则−L^2 的值为_____。

18. VB 提供的标准数据类型，整型类型声明时，其类型关键字是_____；其类型符是_____。

19. 大于 X 的最小整数的 VB 表示形式为_____。

20. 表达式"AB">"CD" Or "x"< "y" And 9 > 3*2 的值是_____。

21. 设 X$ ="abc123456"，则"a"+str$(val(right(X$,4)))的值是_____。
22. 表达式(-3) Mod 8 的值为_____。
23. 计算#10/10/2009#>#11/10/2009#的值是:_____。
24. 设 a=2,b=3,c=4,d=5，则 NOT a<=c OR 4*c=b^2 AND b<> a+c 的值为_____。
25. a 和 b 中有且仅有一个为 1，相应的 VB 逻辑表达式为_____。
26. 表达式(-3) And 8 的值为_____。
27. 对含有多种运算符的表达式，各种运算之间的顺序为_____，算术运算，关系运算，逻辑运算。
28. 表达式 Fix(-32.68) + Int(-23.02) 的值为_____。
29. 把条件 1<=X<12 写成 VB 关系表达式为_____。
30. 在 VB 中，若要求在模块中强制显式地指定变量的数据类型，应当在模块首部的申明部分使用_____语句。
31. VB 6.0 的连接运算符包括_____运算符和_____运算符两种。
32. 若要在字符串 asdgewl 中取 dge，则使用函数_____实现。
33. 在 VB 6.0 中，变量名最长可达_____个字符。
34. 在一条 Dim 语句中可以声明多个变量，如 Dim strVar, intVar, sngVar As Integer，则 strVar、intVar 与 sngVar 的数据类型分别是 Variant、Variant 和_____。
35. 表达式 32\7 MOD 3^2 的值是_____。
36. 在 VB 中，1234、123456&、1.2346E+5、1.2346D+5 这 4 个常数分别表示_____、_____、_____、_____类型。
37. 设 a="Good morning"，语句 Left(a,3) 值为_____，Mid(a,8,4) 值为_____。
38. VB 中的变量按其作用分为全局变量、模块变量、_____。
39. Print "x=" & (2=4) 的结果为_____。
40. A 和 B 同为正整数或同为负整数的 VB 表达式为_____。
41. 如果：I=12:J=3:I=int(-8.6)+I\J+13/3 MOD 5，则 I 值是_____。
42. VB 的运算符包括算术运算符、_____运算符、逻辑运算符和特殊运算符。
43. 设 x 为一个两位数，将 x 个位数和十位数交换后所得两位数 VB 的表达式是_____。
44. Int(-3.5)、Int(3.5)、Fix(-3.5)、Fix(3.5)、Round(-3.5)、Round(3.5)的值分别是_____、_____、_____、_____、_____、_____。
45. 求 x 与 y 之积除以 z 的余数的 VB 表达式为_____。
46. 写出用随机函数产生一个 200～300 整数的 VB 表达式_____。
47. 在窗体上有个命令按钮，然后编写如下事件过程：

    ```
    m=InputBox("enter the first integer")
    n=InputBox("enter the second integer")
    Print n + m
    ```

程序运行后，单击命令按钮，先后在两个输入框中分别输入"1"和"5"，则输出结果为_____。

三、判断题

1. 用 dim 语句声明的局部变量能保存上一次过程调用后的值。（   ）
2. Byte 类型的数据由 2 个字节组成。（   ）

3. 在 VB 中，Dim a,b,c as integer 和 Dim a as integer, b as integer,c as integer 相同。（   ）
4. 在程序运行过程中，变量中的值不会改变，而常量中的值会被改变。（   ）
5. 所有 VB 的变量，都有隐含说明字符和强调声明两种方法来定义。（   ）
6. 在表达式中，运算符两端的数据类型要求一致。（   ）
7. " "是一个字符串，而""不是一个字符串。（   ）
8. 计算机在处理数据时必须将其装入内存，在高级语言中通过内存单元名来访问其中的数据，命名的内存单元就是常量或变量。（   ）
9. 利用 Private Const 声明的符号常量，在代码中不可以再赋值。（   ）
10. Variant 是一种特殊的数据类型，Variant 类型变量可以存储除了定长字符串数据及自定义类型外的所有系统定义类型的数据，Variant 类型变量还可具有 Empty、Error 和 Null 等特殊值。（   ）
11. 在 VB 中，可通过函数 Now 返回计算机系统的日期和时间。（   ）
12. 在 VB 中，函数 Fix(-3.6) 的返回值是−4。（   ）
13. 设 A=3，B=4，C=5，D=6，则表达式 A>B And C<=D Or 2* A>C 的值是 False。（   ）
14. 由变量名对变量的内容进行使用或修改，则使用变量就是引用变量的内容。（   ）
15. 可以用"&"，"+"合并字符串，但是用在变异变量时，"+"可能会将两个数值加起来。
（   ）

第 4 章 程序的控制结构

4.1 知识点

4.1.1 顺序结构

1. [Let]

<变量名> = 表达式

<对象名.属性名> = 表达式

2. Rem 注释内容

'注释内容

3. 暂停语句

Stop

4. 卸载对象语句

Unload <对象名>

5. 结束语句

End

6. 输入框函数

InputBox(<提示信息>[,<标题>][,默认值])

7. 消息框函数

MsgBox(<提示信息>[,<类型值>][,<标题>])

类型值=按钮类型值+图标类型值+默认按钮值

其中，按钮类型值如表 4-1 所示，图标类型值如表 4-2 所示，MsgBox 函数的返回值如表 4-3 所示，默认按钮值如表 4-4 所示。

表 4-1　　　　　　　　　　　　　　　按钮类型值

数值	符号常量值	显示的按钮
0	vbOKOnly	"确定"
1	vbOKCancel	"确定"、"取消"
2	vbAbortRetryIgnore	"终止"、"重试"、"忽略"

续表

数值	符号常量值	显示的按钮
3	VbYesNoCancel	"是""否""取消"
4	vbYesNo	"是""否"
5	vbRetryCancel	"重试""取消"

表 4-2　　　　　　　　　　　　　图标类型值

数值	符号常量值	显示图标
16	vbCritical	停止图标 x
32	vbQuestion	询问图标 ？
48	vbExclamation	警告图标 ！
64	vbInformation	信息图标 i

表 4-3　　　　　　　　　　　　MsgBox 函数的返回值

用户按键	返回符号常量	返回值
确定	vbOK	1
取消	vbCancel	2
终止	vbAbort	3
重试	vbRetry	4
忽略	vbIgnore	5
是	vbYes	6
否	vbNo	7

表 4-4　　　　　　　　　　　　　默认按钮值

数值	符号常量值	默认按钮
0	vbDefaultButton1	第一个
256	vbDefaultButton2	第二个
512	vbDefaultButton3	第三个

4.1.2　选择结构

1. If 单分支

格式 1：

　　If ＜条件＞ Then ＜语句组＞

格式 2：

　　If ＜条件＞ Then
　　　　＜语句组＞
　　End If

说明　　符合条件值为 True 或非 0；不符合条件值为 False 或 0。

2. If 双分支

格式 1：

```
If <条件> Then [<语句组 1>][Else <语句组 2>]
```

格式 2：

```
If <条件> Then
        <语句组 1>
Else
        <语句组 2>
End If
```

3. 分支嵌套使用

```
If <条件 1> Then
    If <条件 2> Then
        <语句组 1>
    Else
        <语句组 2>
    End If
Else
        <语句组 3>
End If
```

4. If 多分支条件语句

```
If  <条件 1>  Then
        [<语句组 1>]
[ElseIf  <条件 2>  Then
        [<语句组 2>]]
…
[ElseIf  <条件 n>  Then
        [<语句组 n>]]
[Else
        [<其他语句组>]]
End If
```

5. Select 多分支选择语句

```
Select  Case  <测试表达式>
        Case  <表达式表 1>
            [<语句组 1>]
        [Case  <表达式表 2>
            [<语句组 2>]]
        ……
        [Case  <表达式表 n>
            [<语句组 n>]]
[Case Else
            [<语句组 n+1>]]
End Select
```

<表达式表>的形式：

```
CASE    3+5,9
CASE    1 TO 9
CASE    IS>10
```

6. IIf 函数

IIf(条件表达式，条件为 True 时的值，条件为 False 时的值)

4.1.3 循环结构

1. Do Loop 循环

（1）Do While——前测型循环：

```
Do   [While <循环条件>]
    <循环体>
Loop
```

（2）Do While——后测型循环：

```
Do
    <循环体>
Loop  [While <循环条件>]
```

（3）Do Until——前测型循环：

```
Do  [Until <条件>]
        <循环体>
Loop
```

（4）Do Until——后测型循环：

```
Do
    <循环体>
Loop [Until <条件>]
```

（5）Exit Do   <循环体>中强制退出循环的选项

2. For…Next 循环

```
For   <循环变量>=<初值>   To   <终值>  [Step <步长>]
        <循环体 1>
        [Exit For]
        <循环体 2>
Next  [<循环变量>]
Exit For         '强制退出循环
```

循环次数：Int((终值-初值)/步长)+1。

4.2 实验内容

4.2.1 顺序结构实验

【实验目的】

1. 掌握顺序结构及其应用。
2. 掌握赋值语句、输入框和输出框的用法。

一、程序改错

第一题

【实验要求】

运行界面如图 4-1 所示,单击"圆的周长"按钮,弹出输入框 inputbox,输入圆的半径,求出圆的周长显示在标签 Label1 中;单击"圆的面积"按钮,弹出输入框 inputbox,输入圆的半径,求出圆的面积显示在标签 Label1 中;单击"退出"按钮,则结束程序运行。

图 4-1 求圆的周长运行界面

【实验步骤】

1. 主要属性设置(见表 4-5)

表 4-5　　　　　　　　　　　对象属性设置

对象	属性	属性值
Label1	AutoSize	True
	Caption	空
	Font	隶书、小三
Command1	Caption	圆的周长
	Font	黑体、五号
Command2	Caption	圆的面积
	Font	黑体、五号
Command3	Caption	退出
	Font	黑体、五号

2. 编写事件代码

```
Private Sub Command1_Click()
    L = InputBox("输入圆的半径", "求圆的周长", "0")
    '**********FOUND**********
    Label1.Caption = "圆的周长为: 2 * 3.14 * Val(L)"
End Sub

Private Sub Command2_Click()
    '**********FOUND**********
    S = InputBox("输入圆的半径", "0", "求圆的面积")
    '**********FOUND**********
    Label1.Caption = "圆的面积为: " & 3.14 * Val(S)
End Sub
```

```
Private Sub Command3_Click()
    End
End Sub
```

第二题

【实验要求】

运行界面如图 4-2 所示，在文本框 Text1、文本框 Text2 和文本框 Text3 中分别输入时间（小时、分、秒），然后单击"计算"按钮，使用消息框输出总计多少秒，单击"结束"按钮，则结束程序运行。

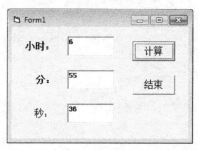 

图 4-2　计算时间运行界面

【实验步骤】

1. 主要属性设置（见表 4-6）

表 4-6　　　　　　　　　　　　对象属性设置

对象	属性	属性值
Label1	AutoSize	True
	Caption	小时：
	Font	宋体、粗体、小四
Label2	AutoSize	True
	Caption	分：
	Font	宋体、粗体、小四
Label3	AutoSize	True
	Caption	秒：
	Font	宋体、粗体、小四
Text1	Text	空
	Font	宋体、粗体、小五
Text2	Text	空
	Font	宋体、粗体、小五
Text3	Text	空
	Font	宋体、粗体、小五
Command1	Caption	计算
	Font	宋体、小四
Command2	Caption	结束
	Font	宋体、小四

2. 编写事件代码

```
Private Sub Command1_Click()
```

```
    Dim hh%, mm%, ss%, Totals!
    Dim Outstr$
    hh = Val(Text1)
    mm = Val(Text2)
    ss = Val(Text3)
'**********FOUND**********
    Totals = hh * 60 + mm * 60 + ss
    Outstr = hh & "小时" & mm & "分" & ss & "秒"
    Outstr = Outstr & vbCrLf & "总计: " & Totals & "秒"
'**********FOUND**********
    MsgBox Totals, , "输出结果"
End Sub
Private Sub Command2_Click()
'**********FOUND**********
    End me
End Sub
```

二、程序填空

第一题

【实验要求】

运行界面如图 4-3 所示,在文本框 Text1 和文本框 Text2 中分别输入不同的数,单击"变量互换"按钮,两个文本框的值互换;单击"退出"按钮,则结束程序运行。

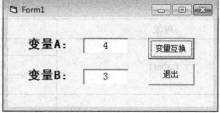

图 4-3 变量互换运行界面

【实验步骤】

1. 主要属性设置(见表 4-7)

表 4-7 对象属性设置

对象	属性	属性值
Label1	AutoSize	True
	Caption	变量 A
	Font	黑体、粗体、四号
Label2	AutoSize	True
	Caption	变量 B
	Font	黑体、粗体、四号
Text1	Text	空
	Alignment	2-Center
	Font	宋体、小四
Text2	Text	空
	Alignment	2-Center
	Font	宋体、小四

对象	属性	属性值
Command1	Caption	变量互换
	Font	宋体、小五
Command2	Caption	退出
	Font	宋体、小五

2. 编写事件代码

```
Private Sub Command1_Click()
'**********SPACE**********
    a =【?】
'**********SPACE**********
    Text1.Text =【?】
'**********SPACE**********
   【?】
End Sub
Private Sub Command2_Click()
    End
End Sub
```

第二题

【实验要求】

运行界面如图 4-4 所示，程序运行后，在标签 Label1 和标签 Label2 中自动随机产生两个 100 以内的数。在文本框 Text1 中输入两数之和（注：输完后按回车键），若输入的答案错误，则弹出消息框，显示"再想想，加油！"，若答案正确，则弹出消息框，显示"恭喜你，答对了！"。单击"下一题"按钮，随机产生两个 100 以内新的加数，同时文本框 Text1 清空。单击"退出"按钮，则结束程序运行。

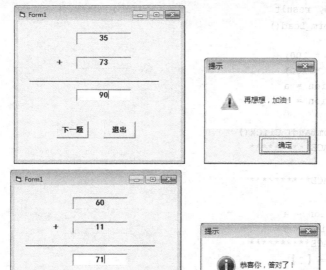

图 4-4　运行界面

【实验步骤】

1. 主要属性设置（见表 4-8）

表 4-8　　　　　　　　　　　　　对象属性设置

对象	属性	属性值
Label1	Alignment	2-Center
	Caption	空
	BorderStyle	1-FixedSingle
	Font	黑体、粗体、小四
Label2	Alignment	2-Center
	Caption	空
	BorderStyle	1-FixedSingle
	Font	黑体、粗体、小四
Label3	AutoSize	True
	Caption	+
	Font	黑体、粗体、三号
Text1	Text	空
	Alignment	2-Center
	Font	黑体、粗体、小四
Command1	Caption	下一题
	Font	黑体、粗体、五号
Command2	Caption	退出
	Font	黑体、粗体、五号

2. 编写事件代码

```
Dim a%, b%, c%, result
Private Sub Form_Load()
  Randomize
  a = Int(Rnd * 100)
  b = Int(Rnd * 100)
  Label1.Caption = a
  Label2.Caption = b
End Sub
Private Sub Command1_Click()
'**********SPACE**********
  a = 【?】
'**********SPACE**********
  b = 【?】
  Label1.Caption = a
  Label2.Caption = b
'**********SPACE**********
  Text1.Text = 【?】
  Text1.SetFocus
End Sub
Private Sub Command2_Click()
  End
End Sub
```

```
Private Sub Text1_KeyPress(KeyAscii As Integer)
   If KeyAscii = 13 Then
     c = Val(Text1.Text)
   '**********SPACE**********
     【?】
     If c = result Then
        MsgBox "恭喜你，答对了！", vbInformation, "提示"
     Else
        MsgBox "再想想，加油！", 48, "提示"
     End If
   End If
End Sub
```

三、编程题

【实验要求】

在文本框 Text1、Text2 和 Text3 中分别输入某学生三门课的成绩，单击"总分"按钮，计算出三门课总成绩并显示在文本框 Text4 中，单击"平均分"按钮，计算出三门课平均成绩并显示在文本框 Text5 中，单击"退出"按钮，结束程序，其运行界面如图 4-5 所示。

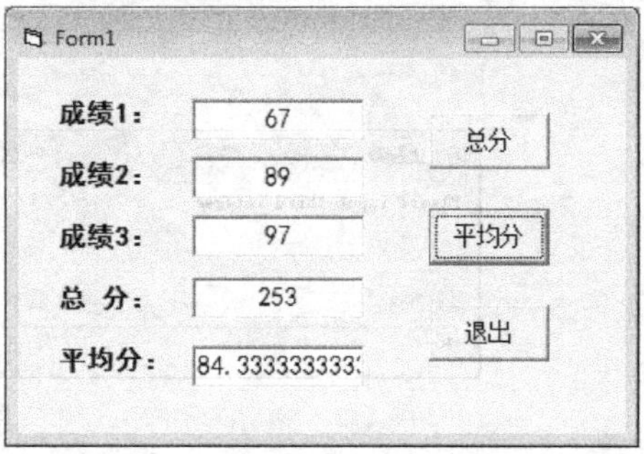

【实验步骤】

1. 界面设计

在窗体上添加 5 个标签（Label11 至 Label5），5 个文本框（Text1 至 Text5），3 个命令按钮（Command1 至 Command3），其大小及位置关系如图 4-5 所示。

图 4-5 运行界面

2. 编写事件代码

```
Private Sub Command1_Click()
'********** Program *********


'********** End *************
End Sub
Private Sub Command2_Click()
'********** Program *********


'********** End *************
End Sub
Private Sub Command3_Click()
'********** Program *********


'********** End *************
End Sub
```

4.2.2 选择结构实验

【实验目的】

1. 掌握 If 条件语句结构及应用。

2. 掌握 Select 多分支选择语句结构及应用。

一、程序改错

第一题

【实验要求】

本程序功能是用鼠标单击窗体时弹出 3 次输入框，依次输入 3 个数，然后将 3 个数由大到小排序，并在窗体上输出。其运行界面如图 4-6 所示。

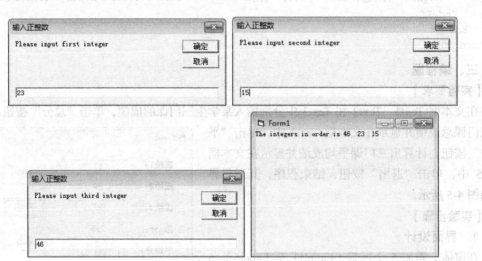

图 4-6　运行界面

【实验步骤】

1. 主要属性设置

无。

2. 编写事件代码

```
Option Explicit
Dim A As Integer
Dim B As Integer
Dim C As Integer
Private Sub Form_Click()
   Dim nTemp As Integer
   A = Val(InputBox("Please input first integer", "输入正整数"))
   B = Val(InputBox("Please input second integer", "输入正整数"))
   C = Val(InputBox("Please input third integer", "输入正整数"))
'**********FOUND**********
   If A <= C Then
      nTemp = A
      A = B
      B = nTemp
   End If
'**********FOUND**********
   If B <= C Then
      nTemp = A
      A = C
      C = nTemp
```

```
      End If
'***********FOUND**********
   If A <= B Then
      nTemp = B
      B = C
      C = nTemp
   End If

   Print "The integers in order is"; A; B; C
End Sub
```

第二题

【实验要求】

某商场为了加速商品流通，采用购物打折的优惠办法，每位顾客一次购物（1）在 100 元及以上者，按九五折优惠；（2）在 200 元及以上者，按九折优惠；（3）300 元及以上者，按八折优惠；（4）500 元及以上者按七折优惠。

运行界面如图 4-7 所示，在文本框 Text1 中输入"购物商品总金额"，然后单击"计算"按钮，文本框 Text2 中将显示打折后的"优惠价"。

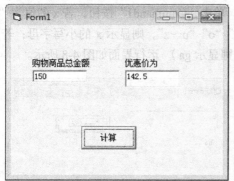 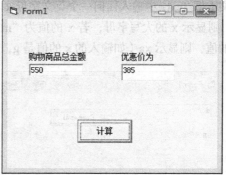

图 4-7 运行界面

【实验步骤】

1. 主要属性设置（见表 4-9）

表 4-9　　　　　　　　　　　对象属性设置

对象	属性	属性值
Label1	Caption	购物商品总金额
Label2	Caption	优惠价为
Text1	Text	空
Text2	Text	空
Command1	Caption	计算

2. 编写事件代码

```
Option Explicit
Private Sub Command1_Click()
   Dim x As Single, y As Single
   x = Val(Text1.Text)
```

```
'**********FOUND**********
    If x > 100 Then
        y = x
    ElseIf x < 200 Then
        y = 0.95 * x
    ElseIf x < 300 Then
        y = 0.9 * x
    ElseIf x < 500 Then
        y = 0.8 * x
    ElseIf x >= 500 Then
        y = 0.7 * x
'**********FOUND**********
    Else
'**********FOUND**********
        Text2.Text = x
End Sub
```

二、程序填空

第一题

【实验要求】

在文本框 Text1 中输入任何一个英文字母 x，单击"Command1"按钮，若 x 的值为"a""c""d~f"，则显示 x 的大写字母；若 x 的值为"m""o""p~z"，则显示 x 的小写字母；若 x 的值为其他的值，则显示 xa（如输入的 x 的值是 g，则显示 ga）。运行界面如图 4-8 所示。

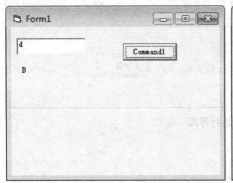 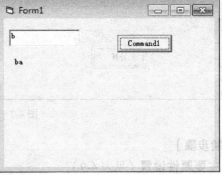

图 4-8 运行界面

【实验步骤】

1. 主要属性设置（见表 4-10）

表 4-10 对象属性设置

对象	属性	属性值
Label1	Caption	空
Text1	Text	空
Command1	Caption	Command1

2. 编写事件代码

```
Private Sub Command1_Click()
x = Text1.Text
```

```
'**********SPACE**********
    【?】
'**********SPACE**********
        Case 【?】
            Label1.Caption = UCase(x)
        Case "m", "o", "p" To "z"
            Label1.Caption = LCase(x)
        Case Else
'**********SPACE**********
            【?】
    End Select
End Sub
```

第二题

【实验要求】

运行界面如图 4-9 所示，设计简易计算器程序，在执行时由用户从键盘上输入两个操作数 A 和 B 及运算符+、-、*、/中的任何一个，然后单击"="命令按钮进行运算，并将运算结果显示在文本框中。

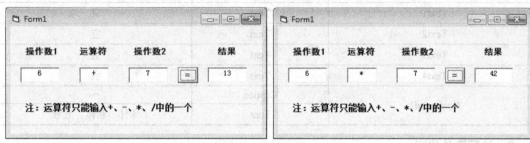

图 4-9　运行界面

在进行除法运算时，若输入的除数是 0，则出现如图 4-10 所示的消息框，并清空除数（操作数 B），将光标定位在该文本框上，以便重新输入除数。

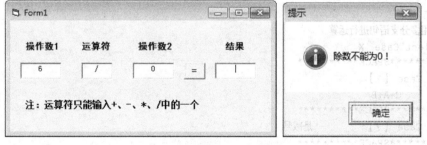

图 4-10　除数为 0 运行界面

【实验步骤】

1. 主要属性设置（见表 4-11）

表 4-11　　　　　　　　　　　　　　对象属性设置

对象	属性	属性值
Label1	AutoSize	True
	Caption	操作数 1
	Font	黑体、粗体、五号

对象	属性	属性值
Label2	AutoSize	True
	Caption	运算符
	Font	黑体、粗体、五号
Label3	AutoSize	True
	Caption	操作数2
	Font	黑体、粗体、五号
Label4	AutoSize	True
	Caption	结果
	Font	黑体、粗体、五号
Label5	AutoSize	True
	Caption	注：运算符只能输入+、-、*、/中的一个
	Font	宋体、粗体、五号
Text1	Text	空
Text2	Text	空
Text3	Text	空
Text4	Text	空
Command1	Caption	=
	Font	宋体、粗体、五号

2. 编写事件代码

```
Private Sub Command1_Click()
   A = Val(Text1.Text)
   B = Val(Text3.Text)
'**********SPACE**********
   X =【?】
   '用多分支语句进行运算
   Select Case X
'**********SPACE**********
     Case 【?】
        C=A+B
'**********SPACE**********
     Case 【?】        '是减号
'**********SPACE**********
        C=【?】
'**********SPACE**********
     Case 【?】        '是乘号
'**********SPACE**********
        C=【?】
'**********SPACE**********
     Case 【?】        '是除号
        If b <> 0 Then
'**********SPACE**********
           C=【?】
```

```
            Else
'**********SPACE**********
            【?】              '用消息框提示错误
                Text3.Text = ""
                Text3.Setfocus
            End If
        End Select
    Text4.Text = C
End Sub
```

三、编程题
【实验要求】

单击窗体，用输入框输入一自然数，判断是"正数""负数"或"零"，并根据输入的数用消息框显示"正数""负数"或"零"。其运行界面如图4-11所示。

图4-11 运行界面

【实验步骤】
1. 界面设计
无。
2. 编写事件代码

```
Private Sub Form_Click()
'********** Program *********




'********** End *************
End Sub
```

4.2.3 循环结构实验

【实验目的】
1. 掌握 Do 循环语句结构及应用。
2. 掌握 For 循环语句结构及应用。

一、程序改错

第一题

【实验要求】
本程序功能是鼠标单击窗体时弹出两次输入框，依次输入两个数 m 和 n，然后求这两个数的最大公因子（最大公约数），并在窗体上输出。其运行界面如图 4-12 所示。

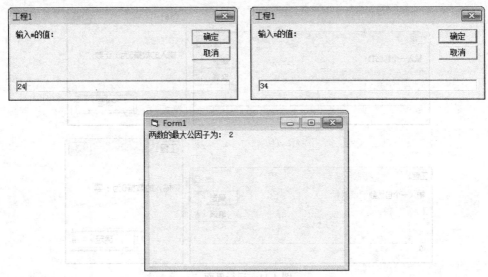

图 4-12 运行界面

【实验步骤】
1. 主要属性设置
无。
2. 编写事件代码
```
Option Explicit
Private Sub Form_Click()
    Dim m As Integer, n As Integer, r As Integer
    m = InputBox("输入 m 的值: ")
    n = InputBox("输入 n 的值: ")
    Do While n <> 0
'**********FOUND**********
        r = m / n
        m = n
'**********FOUND**********
        n = m
    Loop
'**********FOUND**********
```

```
    Print "两数的最大公因子为："; n
End Sub
```

第二题

【实验要求】

用 InputBox 函数输入一个字符串，编写程序按与输入的字符相反的次序用 Msgbox 函数输出这个字符串。如输入字符串为"abcdefgh"，则输出为"hgfedcba"。运行界面如图 4-13 所示。

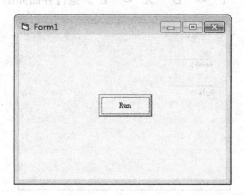

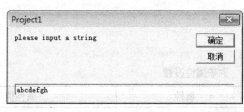

图 4-13 运行界面

【实验步骤】

1. 主要属性设置（见表 4-12）

表 4-12　　　　　　　　　　对象属性设置

对象	属性	属性值
Command1	Caption	Run

2. 编写事件代码

```
Option Explicit
Private Sub Command1_Click()
    Dim pristr As String, outstr As String
    Dim i As Integer
    pristr = InputBox("please input a string")
'**********FOUND**********
    For i = 0 To Len(pristr)
'**********FOUND**********
        outstr = outstr + Mid(pristr, Len(pristr) - i)
    Next i
'**********FOUND**********
    MsgBox outstr, , "The Output Result "
```

```
End Sub
```

二、程序填空

第一题

【实验要求】

本程序将对文本框 txtInput 中输入的字符串中的所有字母进行加密，用鼠标单击窗体，加密结果在文本框 txtCode 中显示。加密方法如下：将每个字母的序号移动 5 个位置，即"A"->"F""a"->"f""B"->"G"…"Y"->"D""Z"->"E"。运行界面如图 4-14 所示。

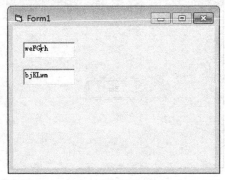

图 4-14 运行界面

【实验步骤】

1. 主要属性设置（见表 4-13）

表 4-13   对象属性设置

对象	属性	属性值
Text1	名称（Name）	txtInput
	Text	空
Text2	名称（Name）	txtCode
	Text	空

2. 编写事件代码

```
Private Sub Form_Click()
    Dim strInput As String * 70    '输入字符串
    Dim Code As String * 70        '加密结果
    Dim strTemp As String * 1      '当前处理的字符
    Dim i As Integer
    Dim Length As Integer          '字符串长度
    Dim iAsc As Integer            '第 i 个字 Ascii 码
'**********SPACE**********
    【?】                          '取字符串
    i = 1
    Code = ""
'**********SPACE**********
    【?】                          '去掉字符串右边的空格，求真正的长度
    Do While (i <= Length)
```

```
'**********SPACE**********
    【 ? 】                    '取第 i 个字符
    If (strTemp >= "A" And strTemp <= "Z") Then
       iAsc = Asc(strTemp) + 5
       If iAsc > Asc("Z") Then iAsc = iAsc - 26
       Code = Left$(Code, i - 1) + Chr$(iAsc)
    ElseIf (strTemp >= "a" And strTemp <= "z") Then
       iAsc = Asc(strTemp) + 5
       If iAsc > Asc("z") Then iAsc = iAsc - 26
       Code = Left$(Code, i - 1) + Chr$(iAsc)
    Else
        Code = Left$(Code, i - 1) + strTemp
    End If
    i = i + 1
  Loop
'**********SPACE**********
    【 ? 】                    '显示加密结果
End Sub
```

第二题

【实验要求】

运行界面如图 4-15 所示，鼠标单击窗体，在窗体上打印出以下图形。

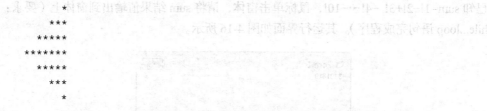

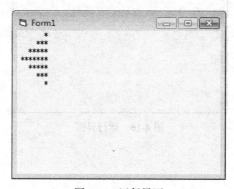

图 4-15　运行界面

【实验步骤】

1. 主要属性设置

无。

2. 编写事件代码

```
Private Sub Form_Click()
   Dim i%, j%, k%
   For i = 3 To 0 Step -1
      For j = 0 To 2 * i
```

```
            Print " ";
         Next j
   '**********SPACE**********
      For k = 5 To 【?】
         Print "*";
      Next k
      Print
   Next i
   For i = 0 To 2
      For j = 0 To 2 * i + 2
   '**********SPACE**********
         【?】
      Next j
      For k = 0 To 4 - 2 * i
         Print "*";
      Next k
   '**********SPACE**********
      【?】
   Next i
End Sub
```

三、编程题
【实验要求】

已知 sum=1!-2!+3! -4!…-10!，鼠标单击窗体，请将 sum 结果值输出到窗体上（要求：使用 do while...loop 语句完成程序）。其运行界面如图 4-16 所示。

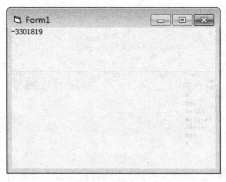

图 4-16 运行界面

【实验步骤】
1. 界面设计
无。
2. 编写事件代码
```
Private Sub Form_Click()
Dim p As Long  'p变量存放阶乘的值
Dim sum As Long
'********** Program *********
```

```
'********** End **************
End Sub
```

4.3 测试题

一、选择题

1. 结构化程序由 3 种基本结构组成，下面不属于 3 种基本结构之一的是（　　）。
   [A] 顺序结构　　　[B] 选择结构　　　[C] 过程结构　　　[D] 循环结构
2. 设有语句：

```
age=Input Box("请输入数值","年龄输入框","25")
```

程序运行后，如果从键盘上输入数值 20，并按〈Enter〉键，则下列叙述中不正确的是（　　）。
   [A] 变量 age 的值是数值 20　　　　　[B] 对话框标题栏中显示的是"年龄输入框"
   [C] "25"为对话框的默认值　　　　　　[D] 对话框的提示文字为"请输入数值"
3. MsgBox 函数返回值的类型是（　　）。
   [A] 整型数值　　[B] 字符串　　[C] 变体　　[D] 数值或字符串
4. 设 a=10，b=5，c=1，执行语句 Printa>b>c 后，窗体上显示的是（　　）。
   [A] True　　[B] False　　[C] 1　　[D] 出错信息
5. 设 a=6，则执行 x=IIf(a>5, -1, 0)后，x 的值为（　　）。
   [A] 5　　[B] 6　　[C] 0　　[D] -1
6. 以下（　　）程序段可以实施 X、Y 变量值的变换。
   [A] Y=X:X=Y　　　　　　　　　　　[B] Z=X:Y=Z:X=Y
   [C] Z=X:X=Y:Y=Z　　　　　　　　　[D] Z=X:W=Y:Y=Z:X=Y
7. 下列多分支选择结构的 Case 语句，写法错误的是（　　）。
   [A] Case 1,5,7,9　　[B] Case 8 To 12　　[C] Case Is < "Man"　　[D] Case 5 To 2
8. VB 中满足 X>8 或者 X≤-8 的正确表达式是（　　）。
   [A] >8　or　X≤-8　　　　　　　　　[B] X>8, X<=-8
   [C] X>8　and　X<=-8　　　　　　　[D] X >8　or　X<=-8
9. 由 For　k=35 to 0 step 3:next k 循环语句控制的循环次数是（　　）。
   [A] 0　　[B] 12　　[C] -11　　[D] -10
10. 选择和循环结构的作用是（　　）。
    [A] 控制程序的流程　　　　　　　　[B] 提高程序的运行速度
    [C] 便于程序的阅读　　　　　　　　[D] 方便程序的调试

二、填空题

1. 要使下列 FOR 语句循环执行 20 次，请在下划线处填入正确的值：

```
For k=_____To -5 Step -2
```

2. 在窗体上有个命令按钮，然后编写如下事件过程：

```
m=InputBox("enter the first integer")
n=InputBox("enter the second integer")
Print n + m
```

程序运行后，单击命令按钮，先后在两个输入框中分别输入"1"和"5"，则输出结果为_____。

3. 当 x=2 时，语句 if x=2 then Print x=2 的结果值是_____。

4. 有如下程序，单击命令按钮，程序运行结果为_____。

```
Private Sub Command1_Click()
   A$ = "A WORKER IS HERE"
   x = Len(A$)
   For I = 1 To x - 1
      B$ = Mid(A$, I, 3)
      If B$ = "WOR" Then S = S + 1
   Next I
   Print S
End Sub
```

5. 执行下列程序段后，输出的结果是_____。

```
For k1=0 To 4
   y=20
   For k2=0 To 3
      y=10
      For k3=0 To 2
         y=y + 10
      Next k3
   Next k2
Next k1
Print y
```

三、判断题

1. For …Next 语句中的初值必须大于终值。（   ）

2. 当 Do While …Loop 或 Do Until…Loop 语句中 While 或 Until 后的表达式的值为 True 或非零时，循环继续。（   ）

3. 用 For 循环写的程序不能用 While 循环来写。（   ）

四、程序填空题

1. 下面的事件过程判断文本框 Text1 中输入的数所在区间，并在文本框 Text2 中输出判断结果。

```
Private Sub Command1_Click()
   Dim int1 As Integer
'**********SPACE**********
   【?】 = Val(Text1.Text)
   Select Case int1
      Case 0
         Text2.Text = "值为0"
'**********SPACE**********
      Case 【?】
         Text2.Text = "值在1和10之间（包括1和10）"
'**********SPACE**********
```

```
            Case Is > 【?】
                Text2.Text = "值大于10"
            Case Else
                Text2.Text = "值小于0"
        End Select
End Sub
```

2. 键盘输入 3 个数,将它们按由大到小的顺序输出,-1 为结束标志。

```
Private Sub Form_Click()
    '**********SPACE**********
    Do While 【?】
        a = Val(InputBox("请输入第一个数:"))
        If a = -1 Then Exit Sub
        b = Val(InputBox("请输入第二个数:"))
        c = Val(InputBox("请输入第三个数:"))
    '**********SPACE**********
        If 【?】 Then t = a: a = b: b = t
        If a < c Then t = a: a = c: c = t
    '**********SPACE**********
        If b < c Then t = b: 【?】
        Print a, b, c
    Loop
End Sub
```

3. 打印出 100~999 之间的所有水仙花数(如果一个数的百位、十位、个位数的立方和等于这个数本身,则这个数为水仙花数)。

```
Private Sub Command1_Click()
    Dim i As Integer, a As Integer, b As Integer, c As Integer
    '**********SPACE**********
    For i = 100 To 【?】
    '**********SPACE**********
        a = Int(i / 【?】)
        b = Int((i - 100 * a) / 10)
    '**********SPACE**********
        c = i - 100 * a - 【?】
        If a * a * a + b * b * b + c * c * c = i Then
            Print i
        End If
    Next i
End Sub
```

五、程序改错题

1. 产生 30 个小于 100 的成绩随机数,统计出优、良、中等、及格、不及格数的个数,并计算出成绩属于优秀段的成绩平均分。

```
Option Explicit
Private Sub Form_Click()
    Dim k%, a%, bjg%, jg%, zd%, lh%, yx As Integer
    Dim pjf As Integer
    Randomize
    pjf = 0
    For k = 1 To 30
```

```
'**********FOUND**********
    a = Int(Rnd())
    print a;
    Select Case a
       Case 0 To 59
          bjg = bjg + 1    '不及格
       Case 60 To 69
          jg = jg + 1      '及格
       Case 70 To 79
          zd = zd + 1      '中等
       Case 80 To 89
          lh = lh + 1      '良好
       Case 90 To 100
          yx = yx + 1      '优秀
'**********FOUND**********
          pjf = pjf + 1
     End Select
  Next k
'**********FOUND**********
  If yx > 0 Then pjf = pjf / 30
  Print "不及格" + Str$(bjg) + "人, 及格" + Str$(jg) + "人, 中等" + Str$(zd) + "人";
  Print "良好" + Str$(lh) + "优秀" + Str$(yx) + "人"
  Print "优秀分数段成绩平均分" & pjf
End Sub
```

2. 以下程序段用于输出 100~300 的所有素数。

```
Option Explicit
Private Sub Form_Click()
   Dim n As Integer, k As Integer, i As Integer, swit As Integer
   For n = 101 To 300 Step 2
     k = Int(Sqr(n))
     i = 2
'**********FOUND**********
     swit = 1
'**********FOUND**********
     While swit = 0
       If n Mod i = 0 Then
         swit = 1
       Else
'**********FOUND**********
         i = i - 1
       End If
     Wend
     If swit = 0 Then
       Print n;
     End If
   Next n
End Sub
```

3. 以下程序段用于打印九九乘法表。

```
Option Explicit
Private Sub Form_Click()
   Dim i As Integer, j As Integer, k As Integer
```

```
    Print Tab(30); "9*9 table"
    Print: Print
    Print "  *  ";
    For i = 1 To 9
'**********FOUND**********
      Print Tab(i * 6); i
    Next i
    Print
    For j = 1 To 9
      Print j; " ";
'**********FOUND**********
      For k = 1 To 9
'**********FOUND**********
        Print Tab(j * 6); j * k; " ";
      Next k
      Print
    Next j
End Sub
```

六、编程题

1. 双击窗体，求出 50 以内所有奇数的和。将结果存入变量 SUM 中。

```
Private Sub Form_dblClick()
    Dim sum As Integer
'******* Program *******




'******* End **********
End Sub
```

2. 双击窗体，求 s=1+1×2+1×2×3+…+1×2×3×…×10，并将结果存到变量 S 中。

```
Private Sub Form_dblClick()
    Dim S As Integer
'******* Program *******




'******* End **********
End Sub
```

第 5 章 数组与过程

5.1 知识点

5.1.1 数组

1. 数组概念

数组是一组相同类型的变量的集合。

数组元素相当于普通变量，默认下标为 0。

2. 静态数组声明

Dim | Private | Public 数组名(下标 1 [，下标 2]…)[As 类型]

下标必须是常数或[下界 to 上界]。

省略[As 类型]，系统认为是变体型数组。

3. 动态数组声明

Dim | Private | Public 数组名()[As 类型]

此时定义的数组的大小是不确定的。

随时用 ReDim 数组名(下标 1[，下标 2]…)语句指定数组的大小。

4. 设定数组下界的默认值

Option Base n

n 是正整数。

5. 使用 Array 函数给一维数组整体赋值

定义一维数组名 A 或 A()类型为 Variant。

使用 Array 函数进行赋值：数组名=Array(值 1，值 2…)

例如，A=Array(0, 1, 2, 3, 4, 5, 6, 7, 8, 9)

6. 控件数组

有相同的对象名称且类型相同的控件，即 Name 属性相同。

有不同的 Index 属性，控件数组的下标用 Index 表示。

一个控件数组有相同的事件过程，在事件过程中用属性 Index 的值来区分是哪个对象触发的事件。例如：

```
Private Sub Command1_Click(Index As Integer)
```

```
        Select  Case  Index
           Case   0
             …
           Case   1
             …
           End   Select
End Sub
```

5.1.2 Sub 子过程

1. 定义格式

[Public | Private | Static]   Sub <过程名>[(形参表)]

   [<语句组>]

End   Sub

<语句组>中可以包含 [Exit   Sub]，其功能是立即退出子过程。

2. 建立子过程
窗体模块子过程：在代码窗口的（通用）处建立，
或打开代码窗口后，选"工具"菜单的"添加过程"选项。
3. 调用子过程
[Call] <过程名> [(<实参表>)]

5.1.3 Function 自定义函数过程

1. 定义格式
[Public | Private | Static] Function <函数名>([<形参表>]) [As <类型>]

   [<语句组>]

   <函数过程名> = <表达式>

End   Function

<语句组>中可以包含 [Exit   Function]，其功能是立即退出函数过程。

2. 调用 Function 过程
<函数过程名> ([<实参表>])

要在命令中调用函数过程，例如：
K=<函数过程名> ([<实参表>])
Print <函数过程名> ([<实参表>])

5.1.4 子过程和函数过程的参数传递

（1）实参：调用过程或函数时的参数，例如：
Call   kk(a, s)
（2）形参：定义过程或函数时的参数，例如：

Sub kk(byval n As Integer,byref　Sum　As Single)

（3）参数的按值传递：形参用 byval 定义，将实参的值传递给形参。

（4）参数的按地址传递：形参用 byref 定义或不用 byval 和 byref 定义。实参和形参共用相同的地址。

（5）在参数传递中，传递的参数是整个数组。

例如，数组名为 A，数组元素是整型，则

实参 A　　：Call　 kk(A)

形参 A()　：Sub　 K3(A()　As　Integer)

（6）在参数传递中，传递的参数是数组元素。

例如，数组名为 A，数组元素是整型，传递的数组元素是 A(2)，则

实参 A(2)　：Call　 kk(A(2))

形参 K　　：Sub　 K4(K　As　Integer)

5.1.5　变量的作用域与生存期

1. 窗体模块的扩展名是.frm。
2. 标准模块的扩展名是.bas。
3. 变量的作用域与生存期如表 5-1 所示。

表 5-1　　　　　　　　　　　　变量的作用域与生存期

变量类型 \ 作用域和生存期	声明的位置	声明的命令动词	能否被本模块的其他过程调用	能否被其他模块调用
局部变量	过程中（事件过程、子过程、函数过程）	Dim 或 Static	不能	不能
模块级变量	通用处（窗体模块或标准模块）	Dim 或 Private	能	不能
全局变量	通用处（窗体模块或标准模块）	Public	能	能（变量名前加窗体名或模块名）

5.2　实验内容

5.2.1　一维数组实验

【实验目的】

1. 掌握一维数组的声明、数组元素的引用。
2. 掌握固定长度数组和动态数组的使用差别。
3. 掌握一维数组常用的操作和常用算法。

一、程序改错

第一题

【实验要求】

输入 10 个正数，自小到大，存于数组 A 中（A（1）～A（10）），编制程序，输入正数 X，

检查它是否存在于 A 中，若存在，显示对应下标；若不存在，则请将 X 插入到 A 中，且不影响 A 中数组序列，运行界面如图 5-1 所示。

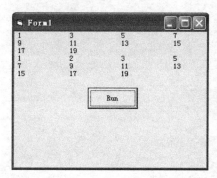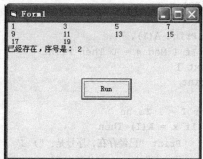

图 5-1　运行界面

【实验步骤】

1. 主要属性设置（见表 5-2）

表 5-2　　　　　　　　　　　　　　对象属性设置

对象	属性	属性值
Command1	Caption	Run

2. 编写事件代码

```
Option Explicit
Private Sub Command1_Click()
Dim x As Integer, I As Integer, nn As Integer, A(11) As Integer, j As Integer
  nn = 10
  I = 1
  While (I <= nn)
    A(I) = Val(InputBox("input number" & "必须大于" & Str(A(I - 1))))
    If A(I) >= A(I - 1) Then
      I = I + 1
    Else
      MsgBox ("请重新输入" & Str(A(I)) & Chr(13) & "必须大于" & Str(A(I - 1)))
    End If
  Wend
  For I = 1 To nn
    Print A(I),
    If I Mod 4 = 0 Then Print
  Next I
  Print
  x = Val(InputBox("Input a Data to Check : "))
  If x < A(1) Then
    For I = nn + 1 To 2 Step -1
      A(I) = A(I - 1)
    Next I
    '**********FOUND**********
    A(I - 1) = x
    For I = 1 To nn + 1
      Print A(I),
      If I Mod 4 = 0 Then Print
    Next I
```

```
            Print
        ElseIf x > A(nn) Then
            '**********FOUND**********
            A(nn) = x
            For I = 1 To nn + 1
                Print A(I),
                If I Mod 4 = 0 Then Print
            Next I
            Print
        Else
            For I = 1 To nn
                If x = A(I) Then
                    Print "已经存在,序号是: "; I
                    Exit Sub
                Else
                    If x > A(I) And x < A(I + 1) Then
                        '**********FOUND**********
                        j = I
                        Exit For
                    End If
                End If
            Next I
            For I = nn + 1 To j + 1 Step -1
                A(I) = A(I - 1)
            Next I
            A(j) = x
            For I = 1 To nn + 1
                Print A(I),
                If I Mod 4 = 0 Then Print
            Next I
            Print
        End If
End Sub
```

第二题

【实验要求】

挑选单数并排序程序：程序启动后由计算机自动产生 20 个属于[100，300]之间的随机整数，单击"显示全部"按钮时，在 Form1 上显示这 20 个随机数；单击"显示奇数"按钮时，在 Form1 上显示其中的奇数；单击"排序"按钮时，在 Form1 上将这些奇数从小到大显示。要求显示格式为每行显示 5 个数据。程序运行结果如图 5-2 所示。

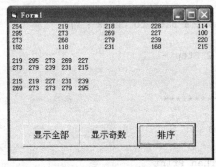

图 5-2　运行结果

【实验步骤】
1. 主要属性设置（见表 5-3）

表 5-3　　　　　　　　　　　　对象属性设置

对象	属性	属性值
Command1	Name	CmdALL
	Caption	显示全部
Command2	Name	Cmdodd
	Caption	显示奇数
Command3	Name	Cmdsort
	Caption	排序

2. 编写事件代码

```
Option Explicit

Private a(20) As Integer, b(20) As Integer
Private k As Integer
Private Sub cmdodd_Click()
Dim I As Integer
k = 0
For I = 1 To 20
    '**********FOUND**********
    If a(I) / 2 = Int(a(I) / 2) Then
        k = k + 1
        b(k) = a(I)
    End If
Next I
For I = 1 To k
    Print b(I);
    '**********FOUND**********
    If Int(I / 5) <> I / 5 Then Print
Next I
Print
End Sub
Private Sub cmdAll_Click()
    Randomize
    Dim I As Integer
    For I = 1 To 20
        '**********FOUND**********
        a(I) = Int(Rnd() * 20 + 100)
        Print a(I),
        If Int(I / 5) = I / 5 Then Print
    Next I
    Print
End Sub
Private Sub cmdsort_Click()
    Dim I As Integer
    Dim J As Integer
    Dim Temp As Integer
    For I = 1 To k - 1
        For J = I To k
```

```
            If b(I) > b(J) Then Temp = b(I): b(I) = b(J): b(J) = Temp
        Next J
    Next I
    For I = 1 To k
        Print b(I);
        If Int(I / 5) = I / 5 Then Print
    Next I
End Sub
```

二、程序填空题

第一题

【实验要求】

输入 n 个数，n 由用户输入。程序的功能是将输入的 n 个数反向输出，程序运行界面如图 5-3 所示。

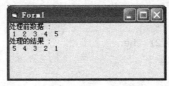

图 5-3 运行界面

【实验步骤】

编写事件代码

```
Private Sub Form_Load()
    Dim n As Integer
    Dim k As Integer
    Dim m As Integer
    Dim h As Integer
    Dim t As Integer
    Show
    Dim a(100) As Single
    n = Val(InputBox("输入个数 n"))
    Print "处理前数据 : "
    '**********SPACE**********
    For k = 1 To 【?】
        a(k) = Val(InputBox("请输入第" & k & "个数"))
        Print a(k);
    Next k
    Print
    '**********SPACE**********
    m = Int(【?】)
    For k = 1 To m
        h = n - k + 1
```

```
    '**********SPACE**********
    t = a(h): 【?】: a(k) = t
  Next k
  Print "处理的结果:"
  For k = 1 To n
    Print a(k);
  Next k
End Sub
```

第二题

【实验要求】

利用自定义类型数组,编写如下程序:模拟实现数据库记录输入、显示与查询功能。程序运行后,单击"新增"按钮,将文本框中的学生信息加到数组中;单击"前一个"或"后一个"按钮,显示当前元素的前或后一个记录;单击"最高"按钮,则显示最高分的记录。并随时显示数组中输入的记录数与当前数组元素的位置。运行界面如图 5-4 所示。

图 5-4 运行界面

【实验步骤】

1. 主要属性设置(见表 5-4)

表 5-4 　　　　　　　　　　对象属性设置

对象	属性	属性值
Label1	Caption	姓名
Label2	Caption	专业
Label3	Caption	总分
Label4	Caption	位置
Label5	Caption	""(空)
Command1	Caption	新增
	Index	0
	Caption	前一个
	Index	1
	Caption	后一个
	Index	2
	Caption	最高
	Index	3

2. 编写事件代码

```
Option Base 1
Private Type StudType
    Name As String * 10          '姓名
    Special As String * 10       '性别
    Total As Single              '总分
End Type
```

```
    Dim n%, i%
    '**********SPACE**********
    Dim stud(1 To 100)  As 【?】

    Private Sub Command1_Click(Index As Integer)
    '**********SPACE**********
    Select Case 【?】
      Case 0      '新增
        If n < 100 Then
          n = n + 1
        Else
          MsgBox "输入人数超过数组声明的个数"
          End
        End If
        i = i + 1
        With stud(n)
        '在stud(n)对象上执行一系列语句到End With结束
          .Name = Text1
          .Special = Text2
          .Total = Val(Text3)
        End With
        Text1 = ""
        Text2 = ""
        Text3 = 0
      Case 1           '前一条
        If i > 1 Then i = i - 1
          With stud(i)
            Text1 = .Name
            Text2 = .Special
            Text3 = .Total
          End With
      Case 2           '后一条
        If i < n Then i = i + 1
          With stud(i)
            Text1 = .Name
            Text2 = .Special
            Text3 = .Total
          End With
      Case 3              '找最高分者
        Max = stud(1).Total
        maxi = 1
        For j = 2 To n
          If stud(j).Total > Max Then
            Max = stud(j).Total
            maxi = j
          End If
        Next j
        With stud(maxi)
          Text1 = .Name
          Text2 = .Special
          Text3 = .Total
        End With
        i = maxi
    End Select
```

```
    Label5 = i & "/" & n    '显示当前位置和总数
End Sub
```

三、编程题

【实验要求】

（事件）双击窗体。

（响应）求 1+2+3+5+8+13+…前 20 项的和，并将结果在窗体上输出。将结果存入变量 SUM 中。

编写事件代码：

```
Private Sub Form_dblClick()
Dim sum As Integer
'*********** Program *************




















'******** End ****************
    YZJ (sum)
End Sub

Private Sub YZJ(i As Integer)

   Dim OUT As Integer
   OUT = FreeFile
   Open App.Path & "\24.out" For Output As #OUT
   Print #OUT, i
   Close #OUT

End Sub
```

5.2.2 二维数组实验

【实验目的】

1. 掌握二维数组的声明、数组元素的引用。
2. 掌握二维数组常用的操作和算法。
3. 掌握控件数组的操作。

一、程序改错

第一题

【实验要求】

以下程序用于建立一个五行五列的矩阵，使其两条对角线上数字为 1，其余位置为 0。结果界面如图 5-5 所示。

【实验步骤】

编写事件代码：

图 5-5 运行界面

```
Option Explicit
Private Sub Form_Click()
Dim x(5, 5), n As Integer, m As Integer
   For n = 1 To 5
     For m = 1 To 5
     '**********FOUND**********
       If n = m Or m + n = 4 Then
          x(n, m) = 1
       Else
          x(n, m) = 0
       End If
     Next m
   Next n
'**********FOUND**********
   For n = 1 To 3
     For m = 1 To 5
'**********FOUND**********
        Print x(m, n)
     Next m
     Print
   Next n
End Sub
```

第二题

【实验要求】

下面的程序用来产生并输出如图 5-6 所示的杨辉三角。

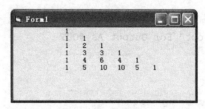

图 5-6 运行界面

【实验步骤】

编写事件代码：

```
Option Explicit
Private Sub Form_Click()
   Dim a(10, 10) As Integer
   Dim i, j ,n As Integer
   n=6
   For i = 1 To n
   '**********FOUND**********
     a(i, n) = 1
     a(i, 1) = 1
   Next i
   For i = 3 To n
   '**********FOUND**********
     For j = 2 To n
       a(i, j) = a(i - 1, j) + a(i - 1, j - 1)
     Next j
```

```
      Next i
      For i = 1 To n
    '**********FOUND**********
        For j = 1 To n
          Print Tab(5 * j + 10); a(i, j);
        Next j
        Print
      Next i
      Print
    End Sub
```

二、程序填空

第一题

【实验要求】

以下程序段用于实现矩阵转置，即将一个 n×m 的矩阵的行和列互换。结果如图 5-7 所示。

图 5-7 运行界面

【实验步骤】

编写事件代码：

```
Private Sub Form_click()
Const n = 3
Const m = 5
Dim a(n, m), b(m, n) As Integer
For I = 1 To n
  '**********SPACE**********
    For j = 1 To 【?】
      a(I, j) = Int(Rnd * 90) + 10
    Next j
Next I
For I = 1 To n
  For j = 1 To m
    b(j, I) = a(I, j)
  Next j
Next I
Print "矩阵转置前"
For I = 1 To n
  For j = 1 To m
    Print a(I, j);
  Next j
'**********SPACE**********
  【?】
Next I
Print "矩阵转置后"
```

```
    For I = 1 To m
      For j = 1 To n
'**********SPACE**********
        Print 【?】
    Next j
      Print
   Next I
End Sub
```

第二题

【实验要求】

使用控件数组实现简易计算器。程序界面如图 5-8 所示。

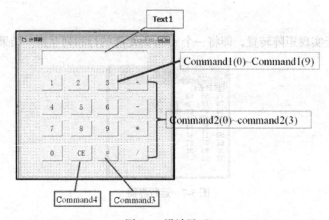

图 5-8 设计界面

【实验步骤】

1. 主要属性设置（见表 5-5）

表 5-5　　　　　　　　　　　　对象属性设置

对象	属性	属性值
Text1	Text	空
Command1	Caption	1, 2, 3~0
	Index	0, 1, 2, 3, ~9
Command2	Caption	+
	Index	0
	Caption	−
	Index	1
	Caption	*
	Index	2
	Caption	/
	Index	3
Command3	Caption	=
Command4	Caption	CE

2. 编写事件代码

```
    Dim op1 As Integer, op2 As Integer, op As String, t As String
```

```
Private Sub Command1_Click(Index As Integer)
'**********SPACE**********
   Text1.Text =【?】
End Sub
Private Sub Command2_Click(Index As Integer)
   op1 = Val(Text1.Text)
   Text1.Text = ""
   op = Command2(Index).Caption
End Sub
Private Sub Command3_Click()
   op2 = Val(Text1.Text)
   '**********SPACE**********
   Select Case 【?】
     Case "+"
        result = op1 + op2
     Case "-"
        result = op1 - op2
     Case "*"
        result = op1 * op2
     Case "/"
        result = op1 / op2
   End Select
   '**********SPACE**********
   Text1.Text =【?】
End Sub

Private Sub Command4_Click()
   Text1.Text = ""
End Sub
```

三、编程题
【实验要求】

求二维数组中最大元素及其所在的行和列并将最大值存入变量 Max 中，将最大值的行位置存入 row 中，列位置存入变量 column 中。

编写事件代码

```
Private Sub Form_Load()
    Show
    Dim a(2, 3) As Integer
    Dim max As Integer, row As Integer, column As Integer
    a(1, 1) = 34: a(1, 2) = 34: a(1, 3) = 43
    a(2, 1) = 34: a(2, 2) = 78: a(2, 3) = 12
'********** Program *********
```

```
'********** End *************
    WWJT max, row, column
End Sub
Private Sub WWJT(x As Integer, y As Integer, z As Integer)
        Dim i As Integer
        Dim s As String
        Dim l As Long
        Dim d As Double
        Dim a(10) As String
        Dim fIn As Integer
        Dim fOut As Integer
        fIn = FreeFile
        fOut = FreeFile
        Open App.Path & "\out.dat" For Output As #fOut
        Print #fOut, x
        Print #fOut, y
        Print #fOut, z
        Close #fIn
        Close #fOut
End Sub
```

5.2.3 Function 过程和 Sub 过程实验

【实验目的】

1. 掌握 Function 过程的定义和调用方法以及变量做函数参数的数据传递。
2. 掌握子过程的定义和调用方法以及数组做函数参数的数据传递。
3. 掌握静态局部变量、局部变量、全局变量的定义和编程技巧。

一、程序改错

第一题

【实验要求】

求 s=2!+4!+6!+8!，阶乘的计算用 Function 过程 fact 实现。设计界面如图 5-9（a）所示，单击窗体后显示运行结果如图 5-9（b）所示，将所给程序段中的错误修改正确并运行该程序。

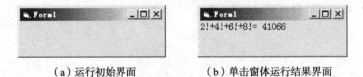

（a）运行初始界面　　　　（b）单击窗体运行结果界面

图 5-9　程序运行界面

【实验步骤】

编写事件代码

```
Option Explicit
Private Sub Form_Click()
```

```
    Dim i As Integer, s As Long
    '**********FOUND**********
    For i = 2 To 8
        s = s + fact(i)
    Next i
    Print "2!+4!+6!+8!=";
    Print s
End Sub
'**********FOUND**********
Public Function fact()
    Dim t As Long
    Dim i As Integer
    t = 1
    For i = 1 To n
        t = t * i
    Next i
    '**********FOUND**********
    fact = i
End Function
```

第二题

【实验要求】

在下列程序段中,过程 fb 可以显示某个数字范围以内的菲波那契数列,现要求通过消息框输入一数字,利用该过程显示不超过该指定数字大小的菲波那契数列元素,运行界面如图 5-10 所示,将所给程序段中的错误修改正确并运行该程序。

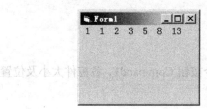 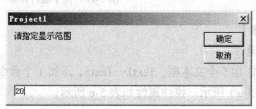

图 5-10 程序运行界面

【实验步骤】

编写事件代码

```
Option Explicit

Public Sub fb(x As Integer)
    Dim i&, j%, k%
    j = 1 : i = 1
    Print i; j;
    k = i + j
    '**********FOUND**********
    Do While k > x
        Print k;
        i = j : j = k
    '**********FOUND**********
        k = i - j
    Loop
End Sub
```

```
Private Sub Form_Click()
    Dim x As Integer
    Cls
    x = Val(InputBox("请指定显示范围"))
    '**********FOUND**********
    call fb y
End Sub
```

二、程序填空

第一题

【实验要求】

设计如图 5-11（a）所示的窗体，程序运行时在 5 个文本框中输入 5 个整数后，单击"求最大值"按钮，求这 5 个数的最大值，运行结果窗口如图 5-11（b）所示，要求调用求 3 个数最大值的函数 max，将所给程序段的【?】处填写完整并运行该程序。

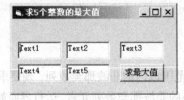 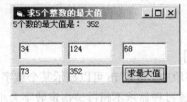

（a）设计界面　　　　　　　　　（b）运行界面

图 5-11　程序运行界面

【实验步骤】

1. 界面设计

在窗体上添加 5 个文本框，Text1…Text5，添加 1 个命令按钮 Command1，各控件大小及位置关系如图 5-11（a）所示。控件属性如表 5-6 所示。

表 5-6　　　　　　　　　　　　　对象属性设置

对　　象	属　　性	属　性　值
Text1、Text2、Text3、Text4、Text5	Text	空
	Alignment	0-Lefe-Justfy
	TabIndex	0、1、2、3、4
Command1	Caption	求最大值
	TabIndex	5

2. 编写事件代码

```
'下面过程 max() 用于求 3 个数中最大值
Public Function max(ByVal a%, ByVal b%, ByVal c%)
'**********SPACE**********
If 【?】 Then
    m = a
Else
    m = b
```

```
        End If
    '**********SPACE**********
        If 【?】 Then
            max = m
        Else
            max = c
        End If
End Function
'利用 max 函数过程求 5 个数中最大值。
Private Sub Command1_Click()
    Dim a%, b%, c%, d%, e%, max1%
    a = Val(Text1.Text)
    b = Val(Text2.Text)
    c = Val(Text3.Text)
    d = Val(Text4.Text)
    e = Val(Text5.Text)
    max1 = max(a, b, c)
    '**********SPACE**********
    max1 = 【?】
    Print "5 个数的最大值是:"; max1
End Sub
```

第二题

【实验要求】

本程序利用二分法查找某数字 n 是否在已排序的数列当中，若在其中则输出其在数列中的位置，否则输出–1。程序运行结果如图 5-12 所示。根据程序功能将所给程序段的【?】处填写完整并运行该程序。

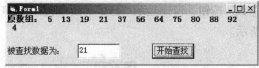

图 5-12　二分法查找数据运行界面

【实验步骤】

1. 界面设计

在窗体上添加 1 个标签控件 Label1，1 个文本框控件 Text1，text 属性值默认为空，1 个命令按钮 Command1，大小及位置关系如图 5-12 所示。

2. 编写事件代码

```
Option Base 1
Sub birsearch(a(), ByVal low%, ByVal high%, ByVal key, index%)
    Dim mid As Integer
    mid = (low + high) \ 2
    If a(mid) = key Then
    '**********SPACE**********
        【?】
        Exit Sub
    ElseIf low > high Then
        index = -1
        Exit Sub
    End If
    If key < a(mid) Then
    '**********SPACE**********
        high = 【?】
    Else
```

```
            low = mid + 1
        End If
        '**********SPACE**********
        Call birsearch(a(), low, high, 【?】, index)
End Sub
Private Sub Command1_Click()
    Dim b() As Variant, index As Integer
    b = Array(5, 13, 19, 21, 37, 56, 64, 75, 80, 88, 92)
    Cls
    Print "原数组:";
    For i = LBound(b) To UBound(b)
        Print b(i);
    Next i
    Print

    n = Val(Text1)
    Call birsearch(b, LBound(b), UBound(b), n, index)
    Print index
End Sub
```

三、编程题

第一题

【实验要求】

已知猴子吃一堆桃子，每天吃桃子总数的一半多一个。到第 n 天时，猴子发现只剩下一个桃子可吃。编写函数 fun，根据 n 值求这堆桃子的数量，例如，n 为 7 时，单击"求这堆桃子总数"按钮，窗体显示"190"，要求使用 Do Until…Loop 语句来实现。程序运行界面如图 5-13 所示。

图 5-13　程序设计界面

【实验步骤】

1. 界面设计

在窗体上添加 1 个标签 Label1，1 个命令按钮 Command1，其大小及位置关系如图 5-13 所示。

2. 编写事件代码

```
Option Explicit
Private Function fun(n As Long) As String
    '**********Program**********



    '********** End **********
End Function
Private Sub Command1_Click()
    Print fun(7)
End Sub
```

第二题

【实验要求】

过程 suixian 可以判断某个数字是否是水仙花数,单击窗口利用该过程找出所有水仙花数。所谓"水仙花数"是指一个 3 位数,其各位数字立方和等于该数本身。运行界面如图 5-14 所示。

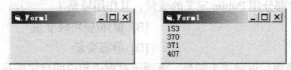

图 5-14 程序运行界面

【实验步骤】

编写事件代码

```
Private Sub Form_Click()
    Dim i As Integer
    For i = 100 To 999
        Call suixian(i)
    Next i
End Sub
Public Sub suixian(x As Integer)
    Dim i%, j%, k%
    '**************** Program **************



    '*************** End ************************
    If i * i * i + j * j * j + k * k * k = x Then Print x
End Sub
```

5.3 测试题

一、选择题

1. 以下属于 VB 中合法的数组元素的是(   )。

   [A] A6   [B] A [6]   [C] A(0)   [D] A{6}

2. 下列叙述中,正确的是(   )。

   [A] 控件数组的每一个成员的 Caption 属性值都必须相同
   [B] 控件数组的每一个成员的 Index 属性值都必须不相同
   [C] 控件数组的每一个成员都执行不同的事件过程
   [D] 对已经建立的多个类型相同的控件,这些控件不能组成控件数组

3. 用语句 Dim A (-3 To 5) As Integer 定义的数组的元素个数是(   )。

   [A] 6   [B] 7   [C] 8   [D] 9

4. 若有数组说明语句为:Dim a(10),则数组 a 包含元素的个数是(   )。

   [A] 10   [B] 22   [C] 8   [D] 11

5. VB 的过程有 3 种，它们是（   ）。

   [A] 事件过程、子过程和函数过程　　[B] Sub 过程、Function 过程和 Property 过程
   [C] 事件过程、函数过程和属性过程　[D] Sub 过程、函数过程和通用过程

6. 在窗体模块的"通用"部分用 Dim 定义的变量，其作用域是（   ）。

   [A] 局部变量　　　　　　　　　　　[B] 窗体/模块级变量
   [C] 全局变量　　　　　　　　　　　[D] 静态变量

7. 在窗体的"通用"部分用 Public 定义的变量，其作用域是（   ）。

   [A] 局部变量　　　　　　　　　　　[B] 窗体/模块级变量
   [C] 全局变量　　　　　　　　　　　[D] 静态变量

8. 要想从子过程调用后返回两个结果，下面子过程语句说明合法的是（   ）。

   [A] Sub f2(ByVal n%, ByVal m%)　　[B] Sub f1(n%, ByVal m%)
   [C] Sub f1(n%,   m%)　　　　　　　[D] Sub f1(ByVal n%,   m%)

9. 下面子过程语句说明合法的是（   ）。

   [A] Sub   f1(ByVal n%( ))　　　　[B] Sub   f1(n%) As Integer
   [C] Function   f1%(f1%)　　　　　[D] Function   f1%(ByVal n%)

10. 下列叙述中错误的是（   ）。

    [A] 子过程以 Sub 开头，以 End Sub 结尾
    [B] 自定义函数以 Function 开头，以 End   Function 结尾
    [C] 只有用 byval 定义形参，该参数传递才是按地址传递
    [D] 只有用 byval 定义形参，该参数传递才是按值传递

11. 单击命令按钮时，下列程序的执行结果为（   ）。

```
Public Sub P (ByVal n As Integer, m As Integer)
   n=n Mod 10
   m=m Mod 10
End Sub
Private Sub Command1_Click()
   Dim x As Integer, y As Integer
   x=8: y=22
   Call P (x, y)
   Print x; y
End Sub
```

   [A] 8 2　　　[B] 8 3　　　[C] 8 22　　　[D] 0 0

12. 单击命令按钮时，下列程序的执行结果为（   ）。

```
Private Sub Command1_Click()
   Dim a As Integer, b As Integer, c As Integer
   a=3: b=4: c=5
   Print P2(c, b, a)
End Sub
Private Function P1(x As Integer, y As Integer, z As Integer)
   P1=2 * x + y + 3 * z
End Function
Private Function P2(x As Integer, y As Integer, z As Integer)
   P2=P1(z, x, y) + x
```

```
    End Function
```
   [A] 21　　　　　[B] 19　　　　　[C] 28　　　　　[D] 34

13. 在窗体上画一个命令按钮，然后编写下列程序：

```
Private Sub Command1_Click()
    Tt 2
End Sub
Sub Tt(a As Integer)
    Static x As Integer
    x=x * a + 1
    Print x;
End Sub
```

连续三次单击命令按钮，输出的结果是（　　　）。

   [A] 1 3 7　　　　[B] 1 4 13　　　　[C] 3 7 4　　　　[D] 2 4 8

二、填空题

1. 若定义一维数组为：Dim a(N To M)，则该数组的元素为_____个。

2. 在 VB 中，若要重新定义一个动态数组的元素个数，应当使用_____语句对其进行重新定义。

3. 要使同一类型控件组成一个控件数组，必须要求_____。

4. 由 Dim a (10) As single 定义的数组占用_____字节的内存空间。

5. 如果在模块的声明段中有 Option Base 0 语句，则在该模块中使用 Dim a(6, 3 To 5)声明的数组有_____个元素。

6. VB 中，变量的声明可分为两种方法：_____数组和动态数组。

7. 在定义有 10 个元素的整型静态数组 a 时，可以定义为_____；也可以在通用部分添加 Option Base 1 语句，并定义为_____。

8. 控件数组的名字由_____属性决定，而数组中每个元素由_____属性指定。

9. 由 Array 函数建立的数组必须是_____类型。

10. 数组元素个数可以改变的数组称为_____。

11. 要获得数组的上界通过_____函数。

12. 在过程调用中，参数的传递可分为两种方式，其中按_____传递方式是默认的。

13. 在过程调用中，参数的传递可分为：地址传递和_____传递两种方式。

14. 在定义子过程或函数的形式参数时，使用关键字_____表示传数值，使用_____表示传地址。

15. 关键字_____声明的局部变量在整个程序运行中一直存在。

16. 传地址方式是当过程被调用时，形参和实参共享_____。

17. 若模块中以关键字 Public 定义子过程，则在_____中都可以调用该过程。

18. 设执行以下程序段时依次输入 5，6，7，执行结果为_____。

```
Option Base 1
Private Sub Command1_Click()
 Dim A(4) As Integer
 For k = 1 To 3
    A(k)=InputBox("""输入数据: "")
 Next k
 Print A(k)
```

19. 设执行以下程序段时依次输入 2，4，6，执行结果为_____。

```
Option Base 1
Dim a(4) As Integer
Dim b(4) AS Integer
For k=0 To 2
  a(k+1)=Val(InputBox(""Enter data: ""))
  b(3 - k) =a(k + 1)
Next k
Print b(k)
```

20. 请用正确内容填空，以下程序的输出结构是：

```
4 7 10
5 8 11
6 9 12
Option Base 1
Private Sub Form_Click()
  Dim i as integer
  Dim j as integer
  Dim a(3, 3) as integer
    For i=1 to 3
      For j=1 to 3
        a(i, j)= _____
        print a(i, j);
      Next j
      print
    Next i
End Sub
```

21. 请用正确的内容填空。下面程序用"选择"法将数组 a 中的 10 个整数按升序排列。

```
Option Base 1
Private Sub Form_Click()
  Dim a
  a = Array(678, 45, 324, 528, 439, 387, 87, 875, 273, 823)
  For i=1 To 9
    For j= i+1 To 10
      If _____ Then
        t=a(i): a(i)=a(j): a(j)=t
      End If
    Next j
  Next i
  For i=1 To 10
    Print a(i);
  Next i
End Sub
```

22. 下列程序运行后，单击窗体，在输入对话框中输入 1、2、3、4、5，输出结果为_____

```
Private Sub Form_Click
  Dim  Arr(10) As  Integer
  For i=6 To 10
    Arr(i)=Inputbox(""请顺序输入1、2、3、4、5"")
  Next i
```

```
            Print Arr(6)+Arr(Arr(6))/ Arr(10)
        End Sub
```

23. 在窗体上画 4 个文本框，并用这四个文本框建立一个控件数组，名称为 Text1（下标从 0 开始，自左至右顺序增大），编写如下事件过程，程序运行后，单击命令按钮，四个文本框中显示的内容分别为_____。

```
        Private Sub Command1_Click()
            For Each txtbox In Text1
                Text1(i) = Text1(i).Index
                i = i + 1
            Next
        End Sub
```

24. 下列程序的运行结果为_____

```
        Private Sub Form_Click()
            Dim a(5) As String
            For i = 1 To 5
                a(i) = Chr(Asc("A") + (i - 1))
            Next i
            For Each b In a
                Print b;
            Next
        End Sub
```

25. 单击一次命令按钮后，下列程序的执行结果为_____

```
        Public Function ZZ(N As Integer)
          Static Sum
          For i=1 To N
            Sum=Sum + i
          Next i
          ZZ=Sum
        End Function
        Private Sub Command1_Click()
          Print ZZ(2) +ZZ(3)
        End Sub
```

26. 下列程序的执行结果为_____

```
        Private Sub Command1_Click()
            Dim s1 As String, s2 As String
            s1="abcdef"
            Call K(s1, s2)
            Print s2
        End Sub
        Private Sub K(ByVal X As String, Y As String)
            Dim T As String
            i=Len(X)
            Do While i >=1
                T=T+ Mid(X, i, 1)
                i=i - 1
            Loop
            Y=T
        End Sub
```

27. 如下程序，运行的结果是_____，函数过程的功能是用辗转相减法求 m、n 的最大公约数。

```
Public Function f(m%, n%)
Do While m<>n
Do While m>n: m=m-n: Loop
Do While n>m: n=n-m: Loop
Loop
F=m
End Function
Private Sub Command1_Click()
Print f(24, 18)
End Sub
```

28. 如下程序，运行的结果是_____，函数过程的功能是用递归函数实现将十进制数 n 以八进制显示。

```
Public Function f(ByVal n%, ByVal r%)
If n<>0 Then
F=f (n\r, r)
Print n Mod r;
End If
End Function

Private Sub Command1_Click()
Print f(100, 8)
End Sub
```

29. 单击一次命令按钮后，下列程序的执行结果是_____。

```
Private Sub Command1_Click()
   s=P(1) + P(2)+ P(3)+ P(4)
   Print s
End Sub
Public Function P(N As Integer)
Static Sum
   For i=1 To N
     Sum=Sum + i
   Next i
   P=Sum
End Function
```

三、判断题

1. 函数过程（Function Procedure）用来完成特定的功能，但不返回相应的结果。（   ）
2. 如果在过程调用时使用按值传递参数，则在被调过程中可以改变实参的值。（   ）
3. 用 Public 申明的变量能被其他模块存取。（   ）
4. 如果没有使用 Public、Private 或者 Friend，Sub 过程在默认情况下是公用的。（   ）
5. 某一过程中的静态变量在过程结束后，静态变量及其值可以在其他过程中使用。（   ）
6. 用表达式作为过程的参数时，使用的是"传地址"方式。（   ）
7. 事件过程与 Sub 过程，它们相同点都是事件驱动，而不同的只是事件过程由控件属性决定，而 Sub 过程是由用户自定义。（   ）
8. 在 Sub 过程中，可以用 Return 语句退出 Sub 过程。（   ）

9. 在过程中用 Dim 和 Static 定义的变量都是局部变量。   (   )
10. 在窗体模块的声明部分中用 Private 声明的变量的有效范围是其所在的工程。   (   )
11. 过程中的静态变量是局部变量,当过程再次被执行时,静态变量的初值是上一次过程调用后的值。   (   )
12. 事件过程由某个用户事件或系统事件触发执行,它不能被其他过程调用。   (   )
13. 在 VB 中将一些通用的过程和函数编写好并封装作为方法供用户直接调用。   (   )
14. 在标准模块的声明部分中用 Publice 声明的变量的有效范围是其所在的工程。   (   )
15. 如果在过程调用时使用按地址传递参数,则在被调过程中不可以改变实参的值。   (   )

四、程序改错题

1. 题目:下面程序段将7个随机整数从小到大排序。

```
Option Explicit
Private Sub Form_Click()
Randomize
Dim t%, m%, n%, w%, a(7) As Integer
For m = 1 To 7
    a(m) = Int(10 + Rnd() * 90)
    Print a(m); " ";
Next m
Print
For n = 1 To 6
    t = n
'**********FOUND**********
    For m = n To 7
'**********FOUND**********
        If a(t) > a(m) Then t = n
    Next m
'**********FOUND**********
    If t = n Then
        w = a(n)
        a(n) = a(t)
        a(t) = w
    End If
Next n
For n = 1 To 7
    Print a(n);
Next n
Print
End Sub
```

2. 题目:移动数组元素,将数组中某个位置的元素移动到指定位置。

```
Option Explicit
Function fMove(a%(), L1%, L2%)
    Dim N As Integer, i As Integer, T As Integer
'**********FOUND**********
    N = LBound(a)
    If L2 = N + 1 Then
'**********FOUND**********
        a(N) = a(L1)
        For i = L1 To N
            a(i) = a(i + 1)
```

```
            Next i
        Else
            T = a(L1)
            a(L1) = a(L2)
            '**********FOUND**********
            a(L1) = T
        End If
End Function

Private Sub Form_Click()
    Dim a%(1 To 5)
    Dim i As Integer
    For i = 1 To 5
        a(i) = i
    Next
    fMove a, 1, 5
    For i = 1 To 5
        Print a(i)
    Next
End Sub
```

五、程序填空题

1. 题目：下面的程序用"冒泡"法将数组 a 中的 10 个整数按从小到大排列，请将程序补充完整。

```
Option Base 1
Private Sub Command1_Click()
Dim a
a = Array(-2, 5, 24, 58, 43, -10, 87, 75, 27, 83)
For i = 10 To 2 Step -1
'**********SPACE**********
For 【?】
'**********SPACE**********
If 【?】 Then
a1 = a(j + 1)
a(j + 1) = a(j)
a(j) = a1
End If
Next j
'**********SPACE**********
【?】
For i = 1 To 10
Print a(i)
Next i
End Sub
```

2. 功能：由筛选法求 m 以内的所有素数。

（1）把 2～m 以内的所有数放入筛中；

（2）找筛中最小的素数，再筛中去掉该素数的所有倍数；

（3）重复（2），直到要找的筛中的最小素数已超出 m 的范围；

（4）在 Pictur1 中打印筛中的素数。

```
Private Sub txtInput_KeyPress(KeyAscii As Integer)
    Dim I As Integer, j As Integer
```

```
            Dim prime(1000) As Integer
            Dim m As Integer, p As Integer
            Dim flag As Boolean
            If KeyAscii = 13 Then
                '**********SPACE**********
                m = Val(【?】)
                For I = 2 To m - 1    '筛子充满数
                    prime(I) = 1
                Next I
                p = 2
                flag = True
                Do
                    Do While p < m And prime(p) = 0  '找筛子中最小的数
                        '**********SPACE**********
                        p = 【?】
                    Loop
                    '**********SPACE**********
                    If p = m Then flag = 【?】  '筛子中的数全求过结束
                    For I = p + p To m - 1 Step p  '在筛子中清除该素数的倍数
                        prime(I) = 0
                    Next I
                    p = p + 1
                Loop While flag = True
                I = 0
                For j = 2 To m - 1  '打印素数
                    '**********SPACE**********
                    If 【?】 Then
                        Picture1.Print j;
                        I = I + 1
                        If I Mod 5 = 0 Then Picture1.Print '一行打印5个素数
                        '**********SPACE**********
                        【?】
                Next j
            End If
        End Sub
```

3. 函数 odd 用于判断一个数是否是奇数。当单击命令按钮时，产生[10, 100]的随机数，调用 odd 过程，判断该数是否为奇数。如果是则显示"奇数"，否则显示"偶数"。

```
        Private Sub odd(n As Integer)
            Print n;
            If n/2<>n\2 Then
                Print "奇数"
            Else
                '**********SPACE**********
                【?】
            End If
        End Sub
        Private Sub Command1_Click()
            Dim x As Integer
            Randomize
            '**********SPACE**********
            x = 【?】
```

4. 两质数的差为2，称此两质数为质数对，下列程序是找出100以内的质数对，并成对显示结果。其中，函数 IsP 判断参数 m 是否为质数。请在程序中下划线处填入相应的内容。

```
Public Function IsP(m)As Boolean
Dim I%
'**********SPACE**********
【?】
For I=2 To Int(Sqr(m))
'**********SPACE**********
If 【?】Then IsP=False
Next I
End Function
Private Sub Command1_Click()
Dim I%
P1=IsP(3)
For I=5 To 100 Step 2
P2=IsP(I)
'**********SPACE**********
If 【?】Then Print I-2, I
'**********SPACE**********
P1【?】
Next I
End Sub
```

5. 在下面的程序段中，函数过程 pd 可以判断任意3个数能否构成三角形的三边，构成三角形时函数 pd 返回值为 True，不构成三角形时函数 pd 返回值为 False，请填空。

```
'**********SPACE**********
Public Function pd(【?】) As Boolean
'**********SPACE**********
If 【?】Then
    pd = True
Else
    pd = False
End If
End Function
```

6. 据统计，人体身高与手印全长存在一定的比例关系：身高（厘米）=手印全长*9.75，编写函数 length 根据人手印长预测人体身高。单击 command1 调用函数。

```
Option Explicit
Private Function length(flong As Single)
    '**********SPACE**********
    【?】 = flong * 9.75
End Function
Private Sub Command1_Click()
    Dim le  As Single
    le = Val(Text1.Text)
    '**********SPACE**********
    h = 【?】
    Text2.Text = h
End Sub
```

第 6 章
Visual Basic 常用控件

6.1 知识点

6.1.1 单选按钮控件

1. 常用属性

Name：名称属性，默认名称 Option1、Option2…
Value：单选按钮是否被选中。值为 False 表示没选中；值为 True 表示选中。
Caption：标题。
Alignment：设置文本对齐方式，设置为 0 表示左对齐，为 1 表示右对齐。
Style 属性：外观样式。设置为 0 表示 Standard 标准模式，设置为 1 表示 Graphical 图形模式，用 Picture 属性设置图片。

2. 主要事件

Click 事件。

6.1.2 复选框控件

1. 常用属性

Name：名称属性，默认名称 Check1、Check2…
Value：复选框是否被选中。为 0 表示未选中，为 1 表示选中，为 2 表示此项禁用。
Name、Caption、Alignment、Style 等属性与单选按钮属性相同。

2. 主要事件

Click 事件。

6.1.3 框架控件

框架控件的常用属性如下。
Enabled：有效属性。值为 True 表示框架内控件有效；值为 False 表示框架内控件无效。
Name、Caption 属性与单选按钮属性相同。

6.1.4 列表框控件

1. 常用属性

Name：默认名称 List1、List2…
List：列表属性，列表框的第 1 项是 List(0)，第 2 项是 List(1)…
Text：列表框中最后被选中列表项的文本值。
ListCount：列表框中列表项的总数。
ListIndex：被选中列表项的索引号（下标），当列表框中没有选择的列表项时，ListIndex 值为-1。若在程序中设置 ListIndex 属性值为 0，表示列表框的第一个选项被选中，列表项呈反相显示。
Sorted：设置排序属性，设置为 True 表示按字母顺序排列；设置为 False 表示按加入先后顺序排列。
Selected：选中属性，逻辑数组 Selected(i)的值为 True 表示第 i+1 项被选中。
Style：设置列表框样式。
[0]默认，传统样式；
[1]有复选框，可同时选择多项。
Columns：设置列属性。值为 0 时所有项目呈单列显示；当值大于或等于 1，项目呈多列显示。
MultiSelect：多重选择属性，有以下 3 种选项。
[0]None：单项选择，禁止选择多项。
[1]Simple：简单多项选择，用鼠标单击选择，或用空格键选择。
[2]Extended：扩展多项选择，选择方法与 Windows 操作方法相同。

2. 常用方法

AddItem：增加列表项。
　　<对象名>.AddItem <成员>[，<下标>]
RemoveItem：删除列表项。
　　<对象名>. RemoveItem <索引>
Clear：删除所有列表项。
　　<对象名>. Clear

3. 主要事件

Click、DblClick。

6.1.5 组合框控件

1. 常用属性

Name：默认名称 Combo1、Combo2…
List、Text、Sorted、ListIndex、ListCount 属性与列表框的相同属性用法相同。
Style：类型属性如表 6-1 所示。

表 6-1　　　　　　　　　　　　　组合框 Style 属性值表

Style 属性值	含义
0	下拉式组合框（允许用户在文本框直接输入文本，列表框隐藏）
1	简单组合框（允许用户在文本框直接输入文本）
2	下拉列表框（不允许用户在文本框直接输入文本，列表框隐藏）

2. 常用方法

AddItem、RemoveItem、Clear 方法与列表框的同名方法用法相同。

3. 主要事件

Click 事件。

6.1.6 图片框控件

1. 常用属性

Name：默认名称 Picture1、Picture2…

Picture：图片属性，可以在属性窗口直接设置，也可以在程序运行时设置。

```
Picture1.Picture=LoadPicture("图形文件路径域及文件名")
Picture1.Picture=LoadPicture()    '清除图片
```

Autosize：值为 True 调整图片框大小，显示整幅图像；值为 False 图片框大小不改变，显示全部或部分图像。

ForeColor：设置前景色。

BackColor：设置背景色。

2. 常用方法

Print　Cls Pset Line Circle。

```
<对象名>. Print <表达式>           '在图片框中输出<表达式>
<对象名>. Cls                      '清除图片框中的数据
<对象名>. Pset [step] (X,Y)[, Color] 画点
<对象名>. Line [[step] (X1,Y1)]-(X2,Y2)[, Color][,B[F]] 画线或矩形
<对象名>. Circle [step] (X,Y),r[, Color[,Start[,End[,aspect]]]] 画圆、椭圆、圆弧和扇形
```

3. 主要事件

Click 事件。

6.1.7 图像框控件

1. 常用属性

Name：默认名称 Image1、Image2…

Picture：用法同图片框控件。

Stretch：值为 True 等比填充，图像自动调整大小以适应图像框；值为 False 全部填充，图像框自动调整大小以适应其中的图像。

2. 常用方法

Move 方法：

　　[<对象名>.]Move　<left>[，[<top>][，[<width>][，<height>]]]

<left>：指示对象左边的水平坐标(x-轴)。

<top>：指示对象顶边的垂直坐标(y-轴)。

<width>：指示对象新的宽度。

<height>：指示对象新的高度。

3. 主要事件

Click 事件。

6.1.8 滚动条控件

1. 常用属性

Name：水平滚动条默认名称 HScroll1…垂直滚动条默认名称 VScroll1…
Min：滚动条的最小值。
Max：滚动条的最大值。
Value：滚动条的当前值。
SmallChange：单击滚动条箭头时滚动条改变量。
LargeChange：单击滚动条内的空白处时滚动条改变量。

2. 主要事件

Change 事件：滚动条 Value 值改变时触发 Change 事件。
Scroll 事件：鼠标拖动滚动滑块时触发该事件。

6.1.9 计时器控件

1. 常用属性

Name：默认名称 Timer1、Timer2…
Interval：设置计时器的定时间隔，单位为 ms，1s=1000ms。
Enabled：计时器是否有效，值为 True 有效，值为 False 无效。

2. 主要事件

Timer 事件：Enabled 值为 True，每隔指定的时间间隔就触发一次 Timer 事件。

6.1.10 直线控件（Line）与形状控件（Shape）

1. 形状控件主要属性

Shape：取值 0～5，分别显示矩形、正方形、椭圆形、圆形、圆角矩形和圆角正方形。
FillStyle：取值 0～7，设置填充模式分别为实心、透明、水平线、垂直线、上对角线、下对角线、交叉线和对角交叉线。
BorderStyle：取值 0～6，边框样式分别为[0]TransParent，透明，边框不可见；[1]Solid，实心边框，最常见；[2]Dash，虚线边框；[3]Dot，点线边框；[4]Dash-Dot，点画线边框；[5]Dash-Dot-Dot，双点画线边框；[6]Inside Solid，内实线边框。
FillColor：设置填充颜色。
BoderWidth：设置边框线宽。
Backcolor：背景颜色。
BackStyle：设置背景是否透明。

2. 直线控件主要属性

BorderStyle 属性与形状控件相同属性用法相同，其他属性略。

6.1.11 文件系统控件

1. 驱动器列表框

常用属性：Drive 属性。
常用方法：Refresh、SetFocus。

主要事件：Change 事件。

2. 目录列表框

常用属性：Path 属性。

常用方法：Refresh、SetFocus。

主要事件：Change 事件。

3. 文件列表框

常用属性：Path、Pattern、FileName、MultiSelect 属性。

常用方法：Refresh、SetFocus。

主要事件：Click、PathChange、PatternChange。

6.2 实验内容

6.2.1 单选按钮、复选框、框架、列表框和组合框控件实验

【实验目的】

1. 掌握单选按钮、复选框、框架、列表框和组合框控件的属性、事件和方法。
2. 掌握单选按钮、复选框、框架、列表框和组合框控件的基本编程方法。

一、程序改错

第一题

【实验要求】

给定年号与月份，判断该年是否闰年，并根据给出的月份来判断是什么季节和该月有多少天？（闰年的条件是：年号能被 4 整除但不能被 100 整除，或者能被 400 整除。）运行界面如图 6-1 所示。

图 6-1 判断季节和天数的运行界面

【实验步骤】

1. 主要属性设置（见表 6-2）

表 6-2　　　　　　　　　　　对象属性设置

对象	属性	属性值
Label1	Caption	年份
Label2	Caption	月份
Label3	Caption	空
	BorderStyle	1—Fixed Single
Text1	Text	空
Combo1	Text	1
	List	1 2 3 ... 12

对象	属性	属性值
Command1	Caption	判断
Frame1	Caption	季节
Option1	Caption	春季
Option2	Caption	夏季
Option3	Caption	秋季
Option4	Caption	冬季

2. 编写事件代码

```
Option Explicit
Private Sub command1_Click()
    Dim Year As Integer, Month As Integer, flag As Integer
    Year = Val(Text1.Text)
    Month = Val(Combo1.Text)
'**********FOUND**********
    If Year Mod 4 = 0 And Year Mod 100 <> 0 And Year Mod 400 <> 0 Then
'**********FOUND**********
        flag = 0
    Else
        flag = 0
    End If
'**********FOUND**********
    Select Case flag
        Case 1
            Label3.Caption = "该月有 31 天"
            Option1.Value = True
        Case 2
            If flag = 1 Then
                Label3.Caption = "该月有 29 天"
                Option1.Value = True
            Else
                Label3.Caption = "该月有 28 天"
                Option1.Value = True
            End If
        Case 3
            Label3.Caption = "该月有 31 天"
            Option1.Value = True
        Case 4
            Label3.Caption = "该月有 30 天"
            Option2.Value = True
        Case 5
            Label3.Caption = "该月有 31 天"
            Option2.Value = True
        Case 6
            Label3.Caption = "该月有 30 天"
            Option2.Value = True
        Case 7
            Label3.Caption = "该月有 31 天"
            Option3.Value = True
```

```
            Case 8
                Label3.Caption = "该月有 31 天"
                Option3.Value = True
            Case 9
                Label3.Caption = "该月有 30 天"
                Option3.Value = True
            Case 10
                Label3.Caption = "该月有 31 天"
                Option4.Value = True
            Case 11
                Label3.Caption = "该月有 30 天"
                Option4.Value = True
            Case 12
                Label3.Caption = "该月有 31 天"
                Option4.Value = True
        End Select
End Sub
```

第二题

【实验要求】

本程序功能用于将学生学过的单词在列表框中显示出来。要完成以下几个功能。

（1）单击"增加单词"按钮，将文本框 Text1 中的单词添加到列表框中并显示列表框中的单词数；

（2）单击"删除单词"按钮，删除列表框中被选中的项并显示列表框中的单词数；

（3）单击"全部清除"按钮，删除列表框的全部选择项并显示列表框中的单词数；

（4）单击"退出"按钮，结束程序。

设计及运行界面如图 6-2 所示，请修正程序中存在的错误。

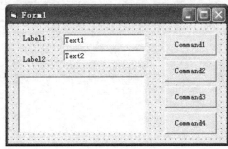
（a）单词管理设计界面

（b）单词管理运行界面

图 6-2 单词管理设计及运行界面

【实验步骤】

1. 界面设计

在窗体上放 2 个标签（Label1 和 Label2）、2 个文本框（Text1 和 Text2）、1 个列表框（List1）及 4 个命令按钮（Command1~Command4）。

2. 编写事件代码

```
Option Explicit
Private Sub Form_Load()
```

```
        Label1.Caption = "请输入单词"
        Label2.Caption = "单词数"
        Command1.Caption = "增加单词"
        Command2.Caption = "删除单词"
        Command3.Caption = "全部清除"
        Command4.Caption = "退出"
        List1.AddItem "apple"
        List1.AddItem "chinese"
        List1.AddItem "child"
        List1.AddItem "double"
        List1.AddItem "park"
        List1.AddItem "top"
        Text1.Text = ""
        Text2.Text = CStr(Me.List1.ListCount)
    End Sub

    Private Sub Command1_click()
        List1.AddItem Text1.Text
        Text2.Text = Str(List1.ListCount)
        Text1 = ""
    End Sub

    Private Sub Command2_click()
        Dim sy As String
        sy = List1.ListIndex
        '**********FOUND**********
        If sy > 0 Then
            '**********FOUND**********
            List1.RemoveItem
            Text2.Text = Str(List1.ListCount)
        End If
    End Sub

    Private Sub Command3_Click()
        '**********FOUND**********
        List1.RemoveItem
        Text2.Text = CStr(List1.ListCount)
    End Sub

    Private Sub Command4_click()
        End
    End Sub
```

二、程序填空

第一题

【实验要求】

在 Text1 中输入一字符串，选中"大写转为小写小写转为大写"项，单击"转换"按钮后，Text2 显示大小写相反的字符串；选中"全部小写"项，单击"转换"按钮后，Text2 显示全部小写的字符串；选中"全部大写"项，单击"转换"按钮后，Text2 显示全部大写的字符串。运行界面如图 6-3 所示。

图 6-3 大小转换运行界面

【实验步骤】
1. 界面设计

在窗体上添加 2 个文本框（Text1 和 Text2）、3 个单选按钮（Option1～Option3）和 1 个命令按钮（Command1），主要属性设置如表 6-3 所示。

表 6-3　　　　　　　　　　　　　　对象属性设置

对象	属性	属性值
Text1 Text2	Text	空
	MultiLine	True
	ScrollBars	2-Vertical
Command1	Caption	转换
Option1	Caption	大写转为小写 小写转为大写
Option2	Caption	全部大写
Option3	Caption	全部小写

2. 编写事件代码

```
Private Sub Command1_Click()
  Dim n As Integer, k As Integer, ch As String, a As String
  '**********SPACE**********
  n = Len(【?】)
  ch = ""
  For k = 1 To n
  '**********SPACE**********
    a = Mid(Text1.Text,【?】,1 )
    If Option1.Value = True Then
      If a >= "a" And a <= "z" Then
        ch = ch + UCase(a)
      ElseIf a >= "A" And a <= "Z" Then
        ch = ch + LCase(a)
      Else
        ch = ch + a
      End If
    End If
    If Option2.Value = True Then
      ch = UCase(Text1)
    End If
    If Option3.Value = True Then
      ch = LCase(Text1)
    '**********SPACE**********
```

```
        【?】
        Text2 = ch
    Next k
End Sub
```

第二题

【实验要求】

其程序的功能是用单选按钮和复选框控制文本框中文字的字体及修饰。要完成如下功能。

(1) 单击"宋体"单选按钮,将文本框中的"今天没下雨"设置为宋体;单击"仿宋_gb2312"单选按钮,将文本框中的"今天没下雨"设置为仿宋体;楷体和黑体同理。

(2) 单击"粗体"复选框,复选框呈选中(加√)状态时,文本框中的"今天没下雨"字体加粗显示;再次单击"粗体"复选框,复选框呈未选中(去掉√)状态时,文本框中的"今天没下雨"字体取消粗体;斜体和下划线同理。

设计及运行界面如图 6-4 所示。

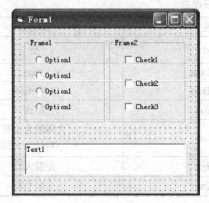
(a) 字体设置设计界面

(b) 字体设置运行界面

图 6-4 字体设置设计及运行界面

【实验步骤】

1. 界面设计

在窗体上添加 2 个框架(Frame1 和 Frame2)和一个文本框(Text1),并在 Frame1 中创建一个控件数组 Option1(0)~Option1(3),在 Frame2 中添加 3 个复选框 Check1、Check2 和 Check3。

2. 编写事件代码

```
Private Sub Form_Load()
    Option1(0).Caption = "宋体"
    Option1(1).Caption = "仿宋_gb2312"
    Option1(2).Caption = "楷体_gb2312"
    Option1(3).Caption = "黑体"
    '**********SPACE**********
    Option1(0).Value=【?】            '默认"宋体"为选中状态
    Frame2.Caption = "修饰"
    Check1.Caption = "粗体"
    Check2.Caption = "斜体"
    Check3.Caption = "下划线"
    Text1.Text = "今天没下雨"
```

```
      Text1.FontSize = 20
End Sub

Private Sub Check1_Click()
   '**********SPACE**********
   If Check1.Value = 【?】 Then
      Text1.FontBold = True
   Else
      Text1.FontBold = False
   End If
End Sub

Private Sub Check2_Click()
   If Check2.Value = 1 Then
      Text1.FontItalic = True
   Else
      Text1.FontItalic = False
   End If
End Sub

Private Sub Check3_Click()
   If Check3.Value = 1 Then
      Text1.FontUnderline = True
   Else
      Text1.FontUnderline = False
   End If
End Sub
Private Sub Option1_Click(Index As Integer)
   '**********SPACE**********
   Text1.FontName = Option1【?】.Caption
End Sub
```

三、编程题

【实验要求】

应用选择法对数组 A 按升序或降序排列，基本思想如下。

（1）对有 n 个数的序列（存放在数组 a(n) 中），从中选出最小（或最大）的数，与第 1 个数交换位置。

（2）除第 1 个数外，其余 n–1 个数中选最小（或最大）的数，与第 2 个数交换位置；

（3）依次类推，选择了 n–1 次后，这个数列已按升序或降序排列。

其设计和运行界面如图 6-5 所示。

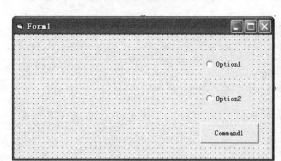　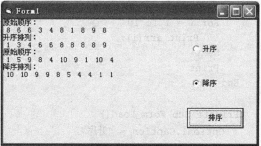

（a）选择排序设计界面　　　　　　　　　　（b）选择排序运行界面

图 6-5　选择排序设计及运行界面

【实验步骤】
1. 界面设计
在窗体上添加 2 个单选按钮（Option1 和 Option2），1 个命令按钮（Command1）。
2. 编写事件代码

```
Option Explicit

Private Sub Sort_Ascending(ByRef a() As Integer, n As Integer)
'********** Program *********




'********** End ************
End Sub
Private Sub Sort_Descending(ByRef a() As Integer, n As Integer)
'********** Program *********




'********** End ************
End Sub
Private Sub Command1_click()
   Show
   Dim i As Integer
   Dim arr(10) As Integer
   Print "原始顺序: "
   For i = 1 To 10
      arr(i) = Int(10 * Rnd + 1)
      Print arr(i);
   Next i
   Print
   If Option1.Value Then
      Sort_Ascending arr, 10
      Print "升序排列: "
   Else
      Print "降序排列: "
      Sort_Descending arr, 10
   End If
   For i = 1 To 10
      Print arr(i);
   Next
   Print
End Sub

Private Sub Form_Load()
   Option1.Caption = "升序"
   Option2.Caption = "降序"
   Command1.Caption = "排序"
End Sub
```

6.2.2 图片框、图像框、计时器和滚动条控件实验

【实验目的】
1. 掌握图片框、图像框、计时器和滚动条控件的属性、事件和方法。
2. 掌握图片框、图像框、计时器和滚动条控件的基本编程方法。

一、程序改错

第一题

【实验要求】
向图片框中输出 100 以内能够被 7 整除的数，要求输出结果为 5 个数一行，如图 6-6 所示。

【实验步骤】

1. 界面设计

在窗体上放 1 个图片框控件 Picture1 和 1 个命令按钮 Command1，属性设置如表 6-4 所示。

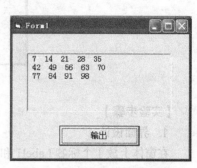

图 6-6 被 7 整除的数运行界面

表 6-4 对象属性设置

对象	属性	属性值
Picture1	Font	宋体 四号字
Command1	Caption	输出

2. 编写事件代码

```
Option Explicit
Private Sub Command1_Click()
   Picture1.Cls
   Dim x As Integer
   Dim i As Integer
'**********FOUND**********
   x = 1
   For i = 1 To 100
      If (i / 7) = (i \ 7) Then
'**********FOUND**********
         Picture1.Print x;
         x = x + 1
         If x Mod 5 = 0 Then Picture1.Print
      End If
'**********FOUND**********
   step x
End Sub
```

第二题

【实验要求】
本程序是一个小动画程序，在窗体上放 1 个标签 Label1 和 1 个计时器控件 Timer1，单击窗体时，每过 1s 标签 Label1 的背景颜色由红到蓝，由蓝到绿，再由绿到红循环变化，并自动修改标签 Label1 的 Left、Top 值使其从左上角沿窗体的对角线移动到窗体的右下角，如此往复从而实现

动画。设计界面如图 6-7 所示。

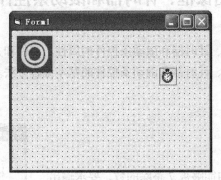

图 6-7 运行的标签设计界面

【实验步骤】
1. 界面设计
在窗体上放 1 个标签 Label1 和 1 个计时器控件 Timer1，属性设置如表 6-5 所示。

表 6-5   对象属性设置

对象	属性	属性值
	Caption	◎
Label1	ForeColor	&H0000FFFF& 黄色
	BackColor	&H000000FF& 红色
Timer1		默认值

2. 编写事件代码

```
Option Explicit
Private flag As Integer
Private Sub Form_click()
   Timer1.Enabled = True
End Sub

Private Sub Form_Load()
   Timer1.Interval = 1000
   Label1.Left = 0
   Label1.Top = 0
   Timer1.Enabled = False
End Sub

Private Sub Timer1_Timer()
   If Label1.Left < Form1. Width And Label1.Top <= Form1. Height Then
      Label1.Left = Label1.Left + 100 * Width / Height
      Label1.Top = Label1.Top + 100 * Width / Height
   Else
      Label1.Left = 0
      Label1.Top = 0
   End If
   If flag = 0 Then
      Label1.BackColor = vbBlue
      '**********FOUND**********
```

```
        flag = 0
    ElseIf flag = 1 Then
        Label1.BackColor = vbGreen
        '**********FOUND**********
        flag = 1
    Else
        Label1.BackColor = vbRed
        '**********FOUND**********
        flag = 2
    End If
End Sub
```

二、程序填空

第一题

【实验要求】

在程序运行时,显示红灯,汽车不动;单击"开始"按钮后,显示绿灯,汽车向右运动;单击右边命令按钮中的一个方向按钮后,则汽车按该按钮上箭头所示的方向移动;单击"停止"按钮,则显示红灯,汽车停止运动。运行界面如图 6-8 所示。

【实验步骤】

1. 界面设计

在窗体上放 3 个图片框(Picture1~Picture3)、2 个命令按钮(Command1~Command2),创建 1 个命令按钮控件数组 Command3(0)~Command3(3)和 1 个计时器(Timer1),属性设置如表 6-6 所示。

图 6-8 汽车运动运行界面

表 6-6　　　　　　　　　　　　　对象属性设置

对象	属性	属性值
Picture1	Picture	TRFFC10C.ICO
	Visible	True
Picture2	Picture	TRFFC10A.ICO
	Visible	False
Picture3	Picture	CARS.ICO
Command1	Caption	开始
Command2	Caption	停止
Command3(0)	Style	1-Graphical
	Picture	ARW04RT.ICO
Command3(1)	Style	1-Graphical
	Picture	ARW04LT.ICO
Command3(2)	Style	1-Graphical
	Picture	ARW04UP.ICO
Command3(3)	Style	1-Graphical
	Picture	ARW04DN.ICO
Timer1	Interval	100
	Enabled	False

2. 编写事件代码

```
Dim m As Integer
Private Sub Command1_Click()
    Picture1.Visible = False
    Picture2.Visible = True
'**********SPACE**********
    Timer1.Enabled =【?】
End Sub

Private Sub Command2_Click()
    Picture2.Visible = False
    Picture1.Visible = True
    Timer1.Enabled = False
End Sub

Private Sub Command3_Click(Index As Integer)
'**********SPACE**********
    m =【?】
End Sub
Private Sub Timer1_Timer()
'**********SPACE**********
    Select Case 【?】
        Case 0: Picture3.Move Picture3.Left + 10
        Case 1: Picture3.Move Picture3.Left - 10
        Case 2: Picture3.Move Picture3.Left, Picture3.Top - 10
        Case 3: Picture3.Move Picture3.Left, Picture3.Top + 10
    End Select
End Sub
```

第二题

【实验要求】

利用一个标签和 3 个水平滚动条，设计一个 RGB 调色板，运行界面如图 6-9 所示。

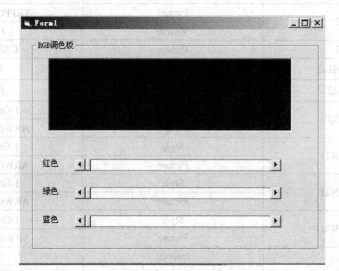

图 6-9 调色板运行界面

【实验步骤】
1. 界面设计

在窗体上放 1 个框架（Frame1）、1 个标签（注意将其更名为 lblColor）、3 个标签（Label1~label3）和 3 个滚动条（注意分别将其更名为 hsRed、hsGreen 和 hsBlue），属性设置如表 6-7 所示。

表 6-7　　　　　　　　　　　　　　对象属性设置

对象	属性	属性值
Frame1	Caption	RGB 调色板
lblColor	BackColor	&H00000000&　黑色
Label1	Caption	红色
Label2	Caption	绿色
Label3	Caption	蓝色

2. 编写事件代码

```
Dim r As Integer, g As Integer, b As Integer
Private Sub Form_Load()
    hsRed.Max = 255
    hsRed.Min = 0
    '**********SPACE**********
    hsGreen.Max = 【?】
    hsGreen.Min = 0
    hsBlue.Max = 255
    hsBlue.Min = 0
    hsRed.SmallChange = 1
    hsGreen.SmallChange = 1
    hsBlue.SmallChange = 1
    hsRed.LargeChange = 5
    hsGreen.LargeChange = 5
    hsBlue.LargeChange = 5
    r = 0
    b = 0
    g = 0
    hsRed.Value = 0
    hsGreen.Value = 0
    hsBlue.Value = 0
    lblColor.BackColor = RGB(r, g, b)
End Sub

Private Sub hsBlue_Change()
'**********SPACE**********
    b = 【?】
    lblColor.BackColor = RGB(r, g, b)
End Sub

Private Sub hsBlue_Scroll()
    b = hsBlue.Value
    lblColor.BackColor = RGB(r, g, b)
End Sub

Private Sub hsGreen_Change()
    g = hsGreen.Value
    lblColor.BackColor = RGB(r, g, b)
```

```
End Sub

Private Sub hsGreen_Scroll()
    g = hsGreen.Value
    lblColor.BackColor = RGB(r, g, b)
End Sub

Private Sub hsRed_Change()
    r = hsRed.Value
    '**********SPACE**********
    lblColor.【?】 = RGB(r, g, b)
End Sub

Private Sub hsRed_Scroll()
    r = hsRed.Value
    lblColor.BackColor = RGB(r, g, b)
End Sub
```

三、编程题

【实验要求】

在窗体上添加一个命令按钮和一个Pictrue控件，请编写Command1_click事件的代码，在Pictrue控件中按图6-10所示格式输出乘法九九表，并存入所给变量SUM中。运行界面如图6-10所示。

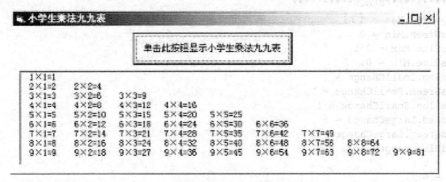

图6-10 显示九九乘法口诀运行界面

【实验步骤】

1. 界面设计

在窗体上放1个图片框（Picture1）、1个命令按钮（Command1），属性设置如表6-8所示。

表6-8　　　　　　　　　　　　　　对象属性设置

对象	属性	属性值
Form1	Caption	小学生乘法九九表
Command1	Caption	单击此按钮显示小学生乘法九九表

2. 编写事件代码

```
Private sub Command1_Click()
    Dim sum As String
    '**************** Program **************
```

```
'*************** End ***************************
    Call YZJ(sum)
End Sub

Private Sub YZJ(i As String)
    Dim OUT As Integer
    OUT = FreeFile
    Open App.Path & "\1.out" For Output As #OUT
    Print #OUT, i
    Close #OUT
End Sub
```

6.2.3 文件系统、直线、形状控件与绘图实验

【实验目的】
1. 掌握文件系统、直线、形状控件的属性、事件和方法。
2. 掌握文件系统、直线、形状控件的基本编程方法。
3. 掌握坐标系的设置及绘图方法。

一、改错题

第一题

【实验要求】

利用图像框和文件系统控件制作一个可以显示位图文件（*.bmp）、JPEG 文件（*.jpg，*.jpeg）和图标文件（*.ico）的图像浏览程序。要求根据驱动器列表框、目录列表框和文件列表框选择磁盘上的图片文件，单击图片文件，图片显示在图像框中。界面设计如图 6-11 所示。

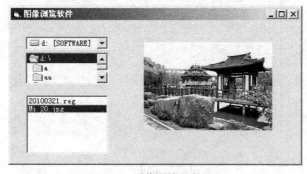

图 6-11 图像浏览程序界面

【实验步骤】
1. 界面设计

在窗体内添加驱动器列表框控件 Drive1、目录列表框控件 Dir1、文件列表框控件 File1 以及图像控件 Image1。属性设置如表 6-9 所示。

表 6-9　　　　　　　　　　　　　　对象属性设置

对象	属性	属性值
Form1	Caption	图像浏览软件
File1	pattern	*.bmp;*.jpg;*.ico
Image1	Stretch	True

2. 编写事件代码

```
Rem 用户改变当前磁盘驱动器时触发的 Change 事件过程如下
Private Sub Drive1_Change()
   '**********FOUND**********
   Dir1.Drive=Drive1.Drive
   Dir1.Refresh
End Sub
Rem 用户改变目录列表框的当前目录时触发的 Change 事件过程如下
Private Sub Dir1_Change()
   '**********FOUND**********
   File1.path= Dir1.Drive
   File1.Refresh
End Sub
Rem  用户单击文件列表框中图像文件时触发的 Click 事件过程如下
Private Sub File1_Click()
   '**********FOUND**********
   Image1.Picture = File1.path & "\" & File1.Filename
End Sub
```

第二题

【实验要求】

已知一个函数 f(x)=1000*sin(x)，利用绘图方法在图片框中显示其图形。结果如图 6-12 所示。

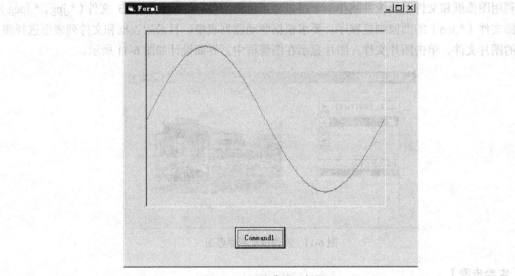

图 6-12 正弦曲线运行界面

【实验步骤】

1. 界面设计

在窗体上放一个图片框（Picture1）和一个命令按钮（Command1），并适当调整图片框大小。

2. 编写事件代码

```
Option Explicit
Private Const pi = 3.14159
Private Sub Command1_Click()
   '**********FOUND**********
```

```
    Dim x As Integer
    Picture1.Scale (-pi, -1200)-(pi, 1200)
    '**********FOUND**********
    For x = -pi To pi Step pi
        '**********FOUND**********
        Picture1.PSet (x, 1000 * pi * Sin(x)), vbRed
    Next x
End Sub
```

二、程序填空

第一题

【实验要求】

用 Circle 方法在图片框正中画出许多不同颜色的同心圆，且不能画出图片框。运行界面如图 6-13 所示。

【实验步骤】

1. 界面设计

在窗体上放一个图片框 Picture1。

2. 编写事件代码

```
Private Sub Picture1_Click()
  Dim CX, CY, Radius, Limit
  Picture1.ScaleMode = 3
  '**********SPACE**********
  CX = 【?】
  CY = Picture1.ScaleHeight / 2
  '**********SPACE**********
  If CX > CY Then Limit = CY Else 【?】
  '**********SPACE**********
  For Radius = 0 To 【?】
      Picture1.Circle (CX, CY), Radius, RGB(Rnd * 255, Rnd * 255, Rnd * 255)
  Next Radius
End Sub
```

第二题

【实验要求】

当程序运行时，单击"开始"按钮，圆半径逐渐变大（圆心位置不变）。当圆充满图片框时则变为红色，并开始逐渐缩小；当缩小到初始大小时又变为蓝色，并再次逐渐变大，如此往复。单击"停止"按钮，则停止变化。运行界面如图 6-14 所示。

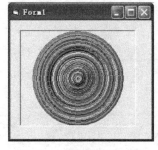

图 6-13　画同心圆运行界面

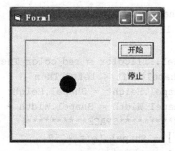

图 6-14　小球动画运行界面

【实验步骤】
1. 界面设计

在窗体上放 1 个图片框（Picture1）、1 个形状控件（Shape1）、1 个计时器（Timer1）和 2 个命令按钮（Command1~Command2），属性设置如表 6-10 所示。

表 6-10　　　　　　　　　　　　　　对象属性设置

对象	属性	属性值
Shape1	Shape	3-Circle
	FillColor	蓝色
Command1	Caption	开始
Command2	Caption	停止
Timer1	InterValue	100
	Enabled	False

2. 编写事件代码

```
Dim left0 As Integer
Const blue_color = &HFF0000, red_color = &HFF&
Private Sub Command1_Click()
  '**********SPACE**********
  Timer1.Enabled = 【?】
End Sub

Private Sub Command2_Click()
  Timer1.Enabled = False
End Sub

Private Sub Form_Load()
  left0 = Shape1.Left
End Sub

Private Sub Timer1_Timer()
  If Shape1.FillColor = blue_color Then
    If Shape1.Left > 0 Then
      Shape1.Height = Shape1.Height + 100
      Shape1.Width = Shape1.Width + 100
      Shape1.Left = Shape1.Left - 50
      Shape1.Top = Shape1.Top - 50
    Else
      '**********SPACE**********
      Shape1.FillColor = 【?】
    End If
  End If
  If Shape1.FillColor = red_color Then
    If Shape1.Left < left0 Then
      Shape1.Height = Shape1.Height - 100
      Shape1.Width = Shape1.Width - 100
      '**********SPACE**********
      【?】 = Shape1.Left + 50
      【?】 = Shape1.Top + 50
```

```
        Else
            '**********SPACE**********
            Shape1.FillColor =【 ? 】
        End If
    End If
End Sub
```

三、编程题

【实验要求】

窗体上有一个圆,相当于一个时钟。当程序运行时通过窗体的 Activate 事件过程在圆上产生 12 个刻度点,并完成其他初始化工作;另有长短两条(红色、蓝色)直线,名称分别为 Line1 和 Line2,表示两个指针。当程序运行时,单击"开始"按钮,则每隔 0.5s Line1(长指针)顺时针转动一个刻度,Line2(短指针)顺时针转动 1/12 个刻度(长指针转动一圈,短指针转动一个刻度),单击"停止"按钮,两个指针停止转动。如图 6-15 所示。

图 6-15 模拟钟表运行界面

【实验步骤】

1. 界面设计

在窗体上放 1 个形状控件(Shape1)、2 个线型控件(Line1、Line2)、1 个计时器(Timer1) 和 2 个命令按钮(Command1、Command2),属性设置如表 6-11 所示。

表 6-11　　　　　　　　　　　　　　对象属性设置

对象	属性	属性值
Shape1	Shape	3-Circle
Line1	BorderColor	&H000000FF&红色
Line2	BorderColor	&H00FF0000&蓝色
Timer1	Interval	500
	Enabled	False
Command1	Caption	开始
Command2	Caption	停止

2. 编写事件代码

```
Const x0 = 1200, y0=1200, radius=1000
Dim a, b, len1, len2
Private Sub Form_Load()
    Shape1.Width=2000
    Shape1.Height=2000
    Shape1.Left=200
    Shape1.Top=200
    Line1.Y1=1200
    Line1.Y2=360
    Line1.X1=1200
    Line1.X2=1200
    Line2.Y1=1200
    Line2.Y2=720
    Line2.X1=1200
    Line2.Y2=1200
End Sub
```

```
Private Sub Command1_Click()
    Timer1.Enabled = True
End Sub

Private Sub Command2_Click()
    Timer1.Enabled = False
End Sub

Private Sub Form_Activate()
    For k=0 To 359 Step 30
        x=radius*Cos(k*3.14159/180+x0
        y=y0-radius*Sin(k*3.14159/180)
        Form1.Circle (x,y),20        '画12个刻度
    Next k
    a=90
    b=90
    len1=Line1.Y1-Line1.Y2
    len2=Line2.Y1-Line2.Y2
End Sub

Private Sub Timer1_Timer()
    a=a-30
    Line1.X2=len1*Cos(a*3.14159/180)+x0
    Line1.Y2=y0-len1*Sin(a*3.14159/180)
    b=b-30/12
    Line2.X2=len2*Cos(b*3.14159/180)+x0
    Line2.Y2=y0-len2*Sin(b*3.14159/180)
End Sub
```

6.3 测试题

一、判断题

1. 单选钮控件和复选框控件都具有 Value 属性，它们的作用完全相同。（   ）

2. *.bmp 格式的图片，如果在 autosize 设为 False 的图片框，它会以图片框大小完整显示出来。（   ）

3. 如果一个列表框一共有 5 个选项，而当选中第三项时，这时列表框的 ListIndex 属性为 2。（   ）

4. 组合框兼有文本框和列表框两者的功能，用户可以通过键入文本或选择列表中的项目来进行选择。（   ）

5. 要使定时器控件起作用，其属性 Interval 不可以设置为 0。（   ）

6. 在框架控件内的几个单选按钮中，只能有一个单选按钮的 Value 属性为 True。（   ）

7. Timer 是时钟控件的唯一事件。（   ）

8. VB 提供的几种标准坐标系统的原点都是在绘图区域的左上角，如果要把坐标原点放在其他位置，则需使用自定义坐标系统。（   ）

9. 对于文件系统控件，当驱动器控件 Drive1 中的驱动器符改变时，文件夹列表控件 Dir1 中显示的文件夹也作相应改变，可以在 Drive1 中的 Change 事件中使用如下命令：Dir1.Path=Drive1.

Drive。(　　)
10. Shape 控件和 Line 控件可以在窗体中移动，因此它们具有 Move 方法。(　　)
11. 滚动条控件可作为用户输入数据的一种方法。(　　)
12. 在图片框中放置的控件既可以在该图片框内移动，也可以移出该图片框外。(　　)
13. 用 Cls 方法能清除窗体或图片框图中用 Picture 属性设置的图形。(　　)
14. 通过改变属性窗口中的 Name 属性，可以改变窗体上显示的标题。(　　)
15. VB 提供的单选按钮的 Value 属性，True 表示选中，False 表示未选中。(　　)
16. 在 VB 中，计时器（Timer）定期激活 Timer 事件，使 Timer 中的代码执行一次。(　　)
17. Line(500,500) – (2500,2500) 命令能够正确画出矩形。(　　)
18. 使用 Print 方法只能在窗体中输出，不能在图片框中输出。(　　)
19. 控件是对象，而窗体不是对象，它只是控件对象的窗口。(　　)
20. 所有的控件在程序运行以后都是可见的。(　　)

二、填空题

1. 列表框中项目的序号是从_____开始的。
2. 在控件之中，_____不能从工具箱中被删除。
3. 如果要每隔 5s 产生一个计时器事件，则 Interval 属性应设置为_____。
4. 要获得列表框 List 数组的元素总个数通过_____属性。
5. 计时器控件只能触发_____事件。
6. 为了在运行时把图形文件 picfile.jpg 装入图片框 Picture1，所使用的语句为_____。
7. _____属性设置为 1，单选按钮和复选框的标题显示在左边。
8. 要改变形状的几何特性，应当在属性窗口中改变的属性是_____。
9. 滚动条产生 Change 事件是因为_____值改变了。
10. 使用 Circle 方法在窗体 Form1 上以 (15,15) 为圆心，10 为半径画圆，具体形式为_____。
11. 设置计时器对象触发事件的时间间隔用_____属性。
12. 一个控件在窗体上的位置由 Top 和_____属性决定，其大小由 Width 和_____属性决定。
13. 滚动条响应的重要事件有_____和 Change。
14. 把 "VB 程序设计" 添加到列表框 lstBooks 的语句为_____。
15. 当用户单击滚动条的空白处时，滑块移动的增量值由_____属性决定。
16. 图像框比图片框占用内存_____，显示速度_____，而且它还可以通过把_____属性设置为 True 来自动调整图像框中图形的大小。
17. 在程序运行时，如果将框架的_____属性设为 False，则框架的标题呈灰色，表示框架内的所有对象均被屏蔽，不允许用户对其进行操作。
18. 在窗体上有两个单选按钮(Option1、Option2)，一个复选框(Check1)。当单选按钮 Option2 和复选框被选中，则 Check1.value 值为_____，Option2.value 值为_____。
19. 将_____属性设置为 1，单选按钮和复选框以图形方式显示。
20. 使用 Scale 方法建立窗体 Form1 的用户坐标系，其中窗体左上角坐标为(-200,250)，右下角坐标为(300, -100)，具体形式为_____。

三、选择题

1. 复选框的 Value 属性为 1 时，表示(　　)。

[A] 复选框未被选中 [B] 复选框被选中
[C] 复选框内有灰色的勾 [D] 复选框操作错误

2. 下面（ ）对象在运行时一定不可见。
   [A] Line [B] Timer [C] Text [D] Option

3. 在 VB 中，下列不能作为存放对象的容器是（ ）。
   [A] 窗体 [B] 框架 [C] 图形框 [D] 图像框

4. 加载 D 盘上 KK 文件夹中的图形文件 JJ.bmp 到图片框 P1 中去的命令是（ ）。
   [A] P1.Picture=Nothing
   [B] P1.Picture= LoadPicture()
   [C] P1.Picture= LoadPicture(D:\KK\JJ.BMP)
   [D] P1.Picture= LoadPicture("D:\KK\JJ.BMP")

5. 当把框架的（ ）属性设置为 False 时，其标题会变灰，框架中所有的对象均被屏蔽？
   [A] Name [B] Enabled [C] Caption [D] Visible

6. 要清除组合框 Combol 中的所有内容，可以使用（ ）语句。
   [A] Combo1.Cls [B] Combo1.Clear
   [C] Combo1.Delete [D] Combo1.Remove

7. 形状控件所显示的图形不可能是（ ）。
   [A] 圆 [B] 椭圆 [C] 圆角正方形 [D] 等边三角形

8. 下列方法中可用于列表框动态添加数据的是（ ）。
   [A] add [B] additem [C] addlist [D] removeitem

9. 引用列表框的最后一项应使用（ ）。
   [A] List1.List(List1.ListCount-1) [B] List1.List(List1.ListCount)
   [C] List1.List(ListCount) [D] List1.List(ListCount-1)

10. 在程序代码中修改滚动条的 value 属性时将激发滚动条的（ ）事件。
    [A] change [B] scroll [C] dragdrop [D] gotfocus

11. 重新定义图片框控件的坐标系统，可采用该图片框的（ ）方法。
    [A] Scale [B] ScaleX [C] ScaleY [D] SetFocus

12. 当在滚动条内拖动滚块时触发（ ）事件。
    [A] KeyUp [B] KeyPress [C] Change [D] Scroll

13. 下列控件中，没有 Caption 属性的是（ ）。
    [A] 框架 [B] 列表框 [C] 复选框 [D] 单选按钮

14. 在文件列表框中设定"文件列表"中显示文件类型应修改该控件的（ ）属性。
    [A] Pattern [B] Path [C] Filename [D] Name

15. 组合框控件是将（ ）组合成一个控件。
    [A] 列表框控件和文本框控件 [B] 标签控件和列表框控件
    [C] 标签控件和文本框控件 [D] 复选框控件和选项按钮控件

16. 在列表框中当前被选中的列表项的序号是由下列哪个属性表示（ ）。
    [A] List [B] Index [C] ListIndex [D] TabIndex

17. VB 6.0 中任何控件都有的属性是（ ）。
    [A] BackColor [B] Caption [C] Name [D] BorderStyle

18. 与 List1.Text 属性值相同的是（　　）。

   [A] List1.ListCount　　　　　　　　[B] List1.List(ListCount-1)

   [C] List1.ListIndex　　　　　　　　[D] List1.List(List.ListIndex)

19. vb 中的坐标系最小刻度为（　　）。

   [A] 缇　　　[B] 像素　　　[C] 厘米　　　[D] 一个标准字符宽度

20. 设组合框 Combo1 中有 3 个项目，则以下能删除最后一项的语句是（　　）。

   [A] Combo1.RemoveItemText　　　　[B] Combo1.RemoveItem 2

   [C] Combo1.RemoveItem 3　　　　　[D] Combo1.RemoveItemCombo1.Listcount

21. 要使某控件在运行时不可显示，应对（　　）属性进行设置。

   [A] Enabled　　[B] Visible　　[C] BackColor　　[D] Caption

22. 选中复选框控件时，value 属性的值可以是（　　）。

   [A] True　　　[B] false　　　[C] 0　　　[D] 1

23. Line(100,100)-Step(400,400)将在窗体（　　）画一直线。

   [A] (200,200)到(400,400)　　　　　[B] (100,100)到(300,300)

   [C] (100,100)到(500,500)　　　　　[D] (100,100)到(400,400)

24. 用来设置粗体字的属性是（　　）。

   [A] FontItalic　　[B] FontName　　[C] FontBold　　[D] FontSize

25. 组合框有 3 种风格，它们由 Style 属性所决定，其中为下拉列表框时，Style 属性值应为（　　）。

   [A] 0　　　[B] 1　　　[C] 2　　　[D] 3

26. 如果在图片框上使用绘图方法绘制一个实心圆，则图片框的（　　）属性决定了该圆的颜色。

   [A] BackColor　　[B] ForeColor　　[C] FillColor　　[D] DrawStyle

27. 要清除已经在图片框 Picture 中打印的字符串而不清除图片框中的图像，应使用语句（　　）。

   [A] P1.Cls　　　　　　　　　　　　[B] P1.picture=LoadPicture("")

   [C] P1.Print ""　　　　　　　　　　[D] P1.piture ""

28. 要获得文件列表框中当前被选中的文件的文件名，则应使用哪个属性（　　）。

   [A] Dir　　　[B] Path　　　[C] Drive　　　[D] FileName

29. 下列可缩放图片的属性是（　　）。

   [A] autosize　　[B] picture　　[C] stretch　　[D] OleDrawMode

30. 在文件列表框的实现文件的多重选择，应修改该控件的（　　）属性。

   [A] filename　　[B] pattern　　[C] path　　[D] multiselect

31. 以下不具有 Picture 属性对象是（　　）。

   [A] 窗体　　[B] 图片框　　[C] 图像框　　[D] 文本框

第 7 章
Visual Basic 高级控件

7.1 知识点

7.1.1 公共对话框控件

公共对话框控件（CommonDialog）提供"打开""另存为""颜色""字体""打印"及"帮助"6 种对话框。

1. 主要属性

Action 属性：取值为 1、2、3、4、5、6，分别向用户提供"打开""另存为""颜色""字体""打印"及"帮助"对话框。例如：

```
Commondialog1.Action=1
```

"打开"与"另存为"对话框主要属性有 Filename、FileTitle、Filter 等。
"颜色"对话框主要属性有 Color、Flags 等。
"字体"对话框主要属性有 Flags、Fontname、Fontsize、FontBold 等。

2. 常用方法

ShowOpen、ShowSave、ShowPrinter、ShowFont、ShowColor、ShowHelp 分别向用户提供"打开""另存为""颜色""字体""打印"及"帮助"对话框。例如：

```
Commondialog1.ShowOpen
```

7.1.2 Windows 公用控件

Windows 公用控件也称 Windows 高级控件，共有 9 个控件，包括图像列表框、工具栏、状态栏、选项卡、进程条、滑块、树视图、列表视图和图像组合控件。

1. 主要属性

图像列表框、工具栏、状态栏、选项卡的属性通常用属性页设置。进程条和滑块主要属性有 Visible、Max（最大值）、Min（最小值）、Value（当前值）和 Orientation（放置方向）等。选项卡属性有 ClientLeft、ClientTop、SelectedItem、Index、Key 等。

2. 主要事件

工具栏的 ButtonClick 事件，滑块的 Scroll 事件和选项卡的 Click 事件。

3. 常用方法

选项卡的 Zorder 方法，树视图的 Add、Clear 方法等。

7.1.3 工具箱中添加"高级控件"选项卡

工具箱中有一个"General"选项卡，上面显示 VB 的所有常用控件，用户也可以在工具箱中添加其他选项卡。例如，在工具箱中添加一个"高级控件"选项卡，在"高级控件"选项卡上再添加工具栏、状态栏等 9 个高级控件。

"高级控件"选项卡的添加方法：在工具箱"General"上的空白处单击鼠标右键，在弹出的快捷菜单中，选择"添加选项卡"选项，在打开的窗口中输入要建立的新选项卡名称"高级控件"，单击"确定"按钮即可在工具箱中看到"高级控件"选项卡。

7.1.4 ActiveX 控件添加到工具箱中

以公共对话框控件添加到工具箱的方法为例，掌握 ActiveX 控件添加到工具箱的步骤。

方法一：在工程菜单里单击"部件"菜单项，弹出"部件"对话框，如图 7-1 所示。在对话框中选择"控件"选项卡，在 ActiveX 控件列表框中查找到"Microsoft CommonDialog Contral 6.0"控件后勾选，单击"应用"按钮或"确定"按钮，公共对话框控件成功添加到工具箱中。

图 7-1 部件对话框

方法二：在工具箱上的空白处单击鼠标右键，弹出的快捷菜单中选择"部件"菜单项，弹出"部件"对话框，后面的操作方法与方法一相同。

7.2 实验内容

【实验目的】
1. 掌握各种 ActiveX 控件添加到工具箱的方法。
2. 掌握利用 CommonDialog 控件设计公共标准对话框。

3. 掌握 ImageList 控件、ToolBar 控件的应用。

一、程序改错

【实验要求】

编写如图 7-2 所示界面，实现如下功能。

（1）当用户单击"打开图片"按钮时，弹出"打开"对话框，从中选定一个图像文件后关闭该对话框，在窗体上显示该文件名，同时在图像框中显示该图片。

（2）单击"设置颜色"按钮，打开颜色对话框，根据颜色对话框的选项，设置文本框内背景颜色。

（3）单击"设置字体"按钮，打开字体对话框，根据字体对话框的选项，设置文本框内文字的格式（字体、字号、字形、颜色、下划线、删除线都设置）。

（4）单击"退出"按钮，关闭窗口。

图 7-2　运行界面

【实验步骤】

1. 主要属性设置（见表 7-1）

表 7-1　　对象属性设置

对象	属性	属性值
Command1	Caption	打开图片
Command2	Caption	设置颜色
Command3	Caption	设置字体
Command4	Caption	退　　出
Image1	Stretch	True
Text1	MultiLine	True
	Text	白日依山尽，黄河入海流。欲穷千里目，更上一层楼。

2. 编写事件代码

```
Private Sub Command1_Click()
    CommonDialog1.ShowOpen
    '**********FOUND**********
    Image1.Picture = LoadPicture(CommonDialog1.Name)
    Print CommonDialog1.FileName
End Sub

Private Sub Command2_Click()
    '**********FOUND**********
    CommonDialog1. ShowOpen
    Text1.BackColor = CommonDialog1.Color
End Sub
```

```
Private Sub Command3_Click()
    CommonDialog1.Flags = &H3 + &H100
    CommonDialog1.ShowFont
    '**********FOUND**********
    Text1.FontName = CommonDialog1.FontSize
    Text1.FontSize = CommonDialog1.FontSize
    Text1.FontBold = CommonDialog1.FontBold
    Text1.FontItalic = CommonDialog1.FontItalic
    Text1.FontUnderline = CommonDialog1.FontUnderline
    Text1.ForeColor = CommonDialog1.Color
    Text1.FontStrikethru = CommonDialog1.FontStrikethru
End Sub

Private Sub Command4_Click()
    Unload Me
End Sub
```

二、程序填空

第一题

【实验要求】

在上一题的基础上设置如图 7-3 所示界面,利用工具栏按钮调整文本框内文字的格式。

图 7-3 运行界面

【实验步骤】

1. 主要属性设置(见表 7-2)

表 7-2 对象属性设置

Button 对象	Index	key	Style
Button1	1	Left	2-tbrButtonGroup
Button2	2	Center	2-tbrButtonGroup
Button3	3	Right	2-tbrButtonGroup
Button4	4		3-tbrSeperator
Button5	5	Bold	1-tbrCheck
Button6	6	Italic	1-tbrCheck
Button7	7	Underline	1-tbrCheck

2. 编写事件代码

```
Private Sub Toolbar1_ButtonClick(ByVal Button As MSComctlLib.Button)
```

```
'**********SPACE**********
Select Case 【?】
  Case 1
    Text1.Alignment = 0
  Case 2
    Text1.Alignment = 2
  Case 3
    Text1.Alignment = 1
    '**********SPACE**********
  Case 【?】
    Text1.FontBold = Not Text1.FontBold
  Case 6
    '**********SPACE**********
    Text1.FontItalic = 【?】
  Case 7
    Text1.FontUnderline = Not Text1.FontUnderline
End Select

End Sub
```

第二题

【实验要求】

在上一题的基础上设置如图 7-4 所示界面，利用状态栏按钮显示图片、文本框内文字的格式、相应的状态信息。

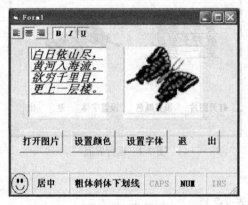

图 7-4 运行界面

【实验步骤】

1. 主要属性设置（见表 7-3）

表 7-3　　　　　　　　　　　　　　对象属性设置

Panels 对象	Index	Alignment	Style	Picture
Panels(1)	1	1-sbrCenter	0-sbrText	图片具体路径
Panels(2)	2	1-sbrCenter	0-sbrText	无
Panels(3)	3	1-sbrCenter	0-sbrText	无
Panels(4)	4	1-sbrCenter	1-sbrCaps	无
Panels(5)	5	1-sbrCenter	2-sbrNum	无
Panels(6)	6	1-sbrCenter	3-sbrIns	无

2. 编写事件代码

```
Private Sub Toolbar1_ButtonClick(ByVal Button As MSComctlLib.Button)
    '**********SPACE**********
    Select Case 【?】
        Case 1
            Text1.Alignment = 0
            StatusBar1.Panels(2).Text = "左对齐"
        Case 2
            Text1.Alignment = 2
            '**********SPACE**********
            StatusBar1.Panels(2).Text = 【?】
        Case 3
            Text1.Alignment = 1
            StatusBar1.Panels(2).Text = "右对齐"
        Case 5
            Text1.FontBold = Not Text1.FontBold
            StatusBar1.Panels(3).Text = StatusBar1.Panels(3).Text + "粗体"
        Case 6
            Text1.FontItalic = Not Text1.FontItalic
            '**********SPACE**********
            StatusBar1.Panels(3).Text = 【?】
        Case 7
            Text1.FontUnderline = Not Text1.FontUnderline
            StatusBar1.Panels(3).Text = StatusBar1.Panels(3).Text + "下划线"
    End Select
End Sub
```

三、编程题

【实验要求】

在上一题的基础上设置如图 7-5 所示界面，利用通用对话框、工具栏按钮及状态栏按钮编辑文本框内的内容。

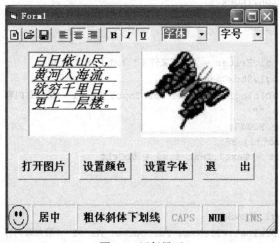

图 7-5　运行界面

【实验步骤】
1. 主要属性设置（见表7-4）

表7-4　　　　　　　　　　　　　　对象属性设置

Button 对象	Index	key	Style
Button1	1	new	0-tbrDefault
Button2	2	open	0-tbrDefault
Button3	3	save	0-tbrDefault
Button4	4		3-tbrSeperator
Button5	5	left	2-tbrButtonGroup
Button6	6	center	2-tbrButtonGroup
Button7	7	right	2-tbrButtonGroup
Button8	8		3-tbrSeperator
Button9	9	Bold	1-tbrCheck
Button10	10	Italic	1-tbrCheck
Button11	11	Underline	1-tbrCheck
Combo1 控件	Text 属性	字体	
Combo1 控件	List 属性	宋体 黑体 楷体_Gb2312 仿宋_Gb2312	
Combo2 控件	Text 属性	字号	
Combo2 控件	List 属性	10 14 18 22	

2. 编写事件代码

参考代码：

```
Private Sub Toolbar1_ButtonClick(ByVal Button As MSComctlLib.Button)
  Select Case Button.Index
    Case 1
      Text1.Text = ""
    Case 2
      CommonDialog1.Filter = "文本文件(*.txt)|*.txt"
      CommonDialog1.ShowOpen
      Open CommonDialog1.FileName For Input As #1  '第九章内容
      Text1.Text = ""
      Do While Not EOF(1)
        Line Input #1, s$
        Text1.Text = Text1.Text & s$ & vbCrLf
      Loop
      Close #1
    Case 3
      CommonDialog1.ShowSave
      Open CommonDialog1.FileName For Output As #2  '第九章内容
      Print #2, Text1.Text
      Close #2
```

```
      Case 5
        Text1.Alignment = 0
        StatusBar1.Panels(2).Text = "左对齐"
      Case 6
        Text1.Alignment = 2
        StatusBar1.Panels(2).Text = "居中"
      Case 7
        Text1.Alignment = 1
        StatusBar1.Panels(2).Text = "右对齐"
      Case 9
        Text1.FontBold = Not Text1.FontBold
        StatusBar1.Panels(3).Text = StatusBar1.Panels(3).Text + "粗体"
      Case 10
        Text1.FontItalic = Not Text1.FontItalic
        StatusBar1.Panels(3).Text = StatusBar1.Panels(3).Text + "斜体"
      Case 11
        Text1.FontUnderline = Not Text1.FontUnderline
        StatusBar1.Panels(3).Text = StatusBar1.Panels(3).Text + "下划线"
    End Select
End Sub
```

7.3 测试题

一、选择题

1. 将通用对话框 CommandDialog1 的类型设置成另存为对话框，可调用该控件的（　　）方法。

   [A] ShowOpen　　　[B] ShowSave　　　[C] ShowColor　　　[D] ShowFont

2. 将 CommonDialog 通用对话框的类型设置为字体对话框，可设置该控件的（　　）属性。

   [A] Font　　　[B] Filter　　　[C] flags　　　[D] Action

3. 下列不能用 commondialog 控件调用的对话框是（　　）。

   [A] open/save　　　[B] help　　　[C] font　　　[D] search

4. 在窗体上画一个的公共对话框 C1 和命令按钮 D1。然后编写如下事件过程：

```
Private Sub D1_Click()
  C1.FileName = ""
  C1.Filter = "all file|*.*|(*.Doc)|*.Doc|(*.Txt)|*.Txt"
  C1.FilterIndex = 2
  C1.DialogTitle = "VBTest"
  C1.Action = 1
End Sub
```

对于这个程序，以下叙述中错误的是（　　）

   [A] 该对话框被设置为"打开"对话框。

   [B] 在该对话框中指定的默认文件名为空。

   [C] 该对话框的标题为 VBTest。

   [D] 在该对话框中指定的默认文件类型为文本文件（*.Txt）。

5. 窗体上有一个名称为 Cd1 的公共对话框控件和由四个命令按钮组成的控件数组 Cmd1，其

index 属性值从左到右分别为 0、1、2、3，窗体界面如图 7-6 所示。命令按钮的事件过程如下：

图 7-6　公共对话框

```
Private Sub Cmd1_Click(Index As Integer)
 Select Case Index
     Case 0: Cd1.Action=1
     Case 1: Cd1.ShowSave
     Case 2: Cd1.Action=5
     Case 3: End
   End Select
End Sub
```

对上述程序，下列描述中错误的是（　　）

[A] 单击"打开"按钮，显示打开文件的对话框。

[B] 单击"保存"按钮，显示保存文件的对话框。

[C] 单击"打印"按钮，能够设置打印选项，并执行打印操作。

[D] 单击"退出"按钮，结束程序的运行。

6. 设已经在窗体上添加一个公共对话框控件 C1，以下正确的语句是（　　）

[A] C1.Filter=All file|*.*|位图文件(*.Bmp)|*.Bmp

[B] C1.Filter="All file"|*.*|"位图文件(*.Bmp)"|*.Bmp

[C] C1.Filter={All file|*.*|位图文件(*.Bmp)|*.Bmp}

[D] C1.Filter="All file|*.*|位图文件(*.Bmp)|*.Bmp"

二、填空题

1. 在控件之中，_____单独保存在，OCX 文件中，在必要时可以加入到工具箱中。

2. 一个工程可以包括多种类型的文件，其中 ActiveX 控件的文件扩展名为_____。

3. 假定有一个通用对话框控件 CommonDialogl，除了用 CommonDialog1.Action=3 显示颜色对话框之外，还可以用_____方法显示。

4. 窗体中有一公共对话框 Comdialog1 和一个命令按钮 Command1，当单击按钮时打开字体对话框。请将程序补充完整。

```
Private Sub Command1_Click()
   ComDialog1._____
End Sub
```

5. 通用对话框可以提供 6 种形式的对话框,通过设置_____属性值或调用 Show 方法来建立不同类型的对话框。

6. 在打开文件对话框时，选定位于"c:\ab\cd"目录下的"ab.txt"的文件，则文件对话框的 FileName 属性值为_____，FileTitle 属性值为_____。

7. 在窗体内有一个命令按钮 Cmd1 和一个公共对话框 C1，命令按钮的事件代码如下：

```
Private Sub Cmd1_Click()
   C1.FileName = ""
   C1.Filter="所有文件(*.*)|*.*|Word 文档(*.Doc)|*.Doc|文本文件(*.Txt)|*.Txt"
```

```
        C1.FilterIndex = 2
        C1.DialogTitle = "OpenFile"
        C1.Action = 1
End Sub
```

则文件对话框的标题是_____。文件对话框中"文件类型"框中显示的是_____。单击"文件类型"框右端的箭头,下拉框中显示的是_____。

8. 使用公共对话框控件打开字体对话框时,如果要在字体对话框中列出可用的屏幕字体和打印机字体,必须设置对话框控件的 Flags 属性值为_____。

9. 要使用通用对话框控件,必须首先将其添加到工具箱中,需要在"部件"对话框"控件"选项卡的控件列表中选择_____控件,然后单击确定命令按钮。

10. 与语句 CommonDialog1.Action=2 等价的语句是_____。

三、判断题

1. ActiveX 控件是扩展名为*.ocx 的独立文件,使用时需用"工程"/"部件"载入或移去。 (   )

2. Circle 方法绘制扇形或圆弧图形时,图形的形状不仅与起始角、终止角的大小相关,而且与起始角、终止角的正或负相关。 (   )

3. 若已在窗体中加入了一个通用对话框:要求在运行时,通过 ShowOpen 打开对话框时,只显示扩展名为 DOC 的文件,则对通用对话框的 Filter 的属性设置应该是:""(*.DOC)|(.DOC)""。
(   )

4. 通用对话框只能用 SHOW 方法进行调用。 (   )

5. 在 VB 6.0 中,如果要增加工具箱中的控件,应执行 VB "文件"菜单中的命令。 (   )

6. 在 VB 的工具栏中包括了所有的 VB 控件,我们不能再加载其他的控件。 (   )

7. 在利用通用对话框件显示字体对话框之前必须设置 Flags 属性,否则将发生不存在字体的错误。 (   )

8. ActiveX 部件是可以重复使用的编程代码和数据。 (   )

第8章 菜单

8.1 知识点

8.1.1 菜单的组成

菜单分为下拉式菜单和弹出式菜单，下拉式菜单由菜单标题组成，每个菜单标题下包括若干个菜单项，分隔条用来将菜单项分组。右侧含有指向三角符号的菜单项，是子菜单标题，鼠标经过右侧指向三角符号弹出下一级菜单；选择含有省略号"…"的菜单项会弹出一个对话框；灰色菜单项表示执行该菜单项的条件不具备，禁止执行。使用频率高的菜单项通常设有快捷键和访问键。下拉式菜单的组成如图8-1所示。

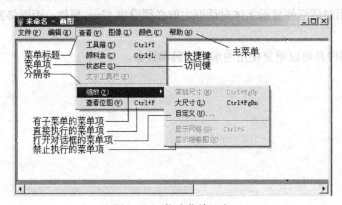

图8-1 下拉式菜单组成

8.1.2 菜单编辑器

VB 6.0 环境中，使用菜单编辑器可以快捷地设计制作出用户所需的菜单界面，但菜单项的功能需要编程来实现。程序代码写在菜单项的 Click 事件中。

菜单编辑器的打开方法有3种。

方法一：在设计菜单的窗体内单击鼠标右键，在弹出的快捷菜单中选择"菜单编辑器"选项即可。

方法二：打开设计菜单的窗体，单击"工具"菜单下的"菜单编辑器"命令。

方法三：打开设计菜单的窗体，直接单击常用工具栏上的"菜单编辑器"按钮 。

菜单编辑器如图 8-2 所示，已建有 7 个菜单项，在菜单编辑器内利用上下箭头按钮调整菜单项的前后顺序，利用左右箭头按钮调整菜单项的级别。

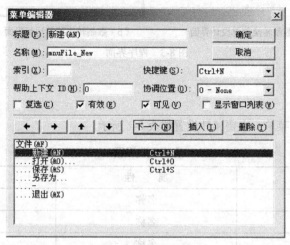

图 8-2　菜单编辑器

8.1.3　下拉式菜单

（1）菜单设计：用菜单编辑器设计菜单及属性。
（2）属性：标题 Caption、名称 Name、可见 Visible、有效 Enabled、复选 Checked 等属性。
（3）事件：菜单项的 Click 事件。

8.1.4　弹出式菜单

（1）菜单设计：用菜单编辑器设计菜单及属性。
（2）属性：标题 Caption、名称 Name、可见 Visible、有效 Enabled、复选 Checked 等属性。
（3）方法：PopupMenu 方法。

8.2　实验内容

菜单实验

【实验目的】
1. 掌握用菜单编辑器设计菜单的方法。
2. 掌握菜单程序的设计方法。

一、改错题

【实验要求】

设计一个如图 8-3（a）所示的菜单，分别用来将八进制数转换成十进制数和将十进制数转换成二进制数，运行界面如图 8-3（b）所示。

（a）数制转换菜单设计界面

（b）数制转换菜单运行界面

图 8-3　程序界面

【实验步骤】

1. 界面设计

用菜单编辑器设计菜单，主要属性设置如表 8-1 所示。

表 8-1　　　　　　　　　　　　对象属性设置

菜单项对象	属　　性	属　性　值
数制转换(&C)	名称	mnu_Change
…八==>十	名称	mnu_OtoD
…十==>二	名称	mnu_DtoB
退出（&E）	名称	mnu_Exit

2. 编写事件代码

```
Option Explicit
Private Sub mnu_OtoD_Click()
   Dim oct_num As String
   Dim length As Integer, s As Integer, n As Integer, i As Integer, d As String
   oct_num = InputBox("请输入一八进制数")
'**********FOUND**********
   length = length(oct_num)
   s = 0
   n = 0
'**********FOUND**********
   For i = length To 1
      d = Mid(oct_num, i, 1)
      s = s + d * 8 ^ n
      n = n + 1
'**********FOUND**********
   Loop
   Print oct_num ;"对应的十进制数是:"; s
End Sub

Private Sub mnu_DtoB_Click()
   Const n = 8
   Dim a(n)   As Integer, s As String, m As Integer, x As Integer
   x = Val(InputBox("请输入一个 0～255 的正整数："))
   Print x; "对应的二进制数是: ";
'**********FOUND**********
   For m = 1 To n
      a(m) = x Mod 2
'**********FOUND**********
      x = x / 2
   Next m
```

```
      s = " "
      For m = n To 0 Step -1
'**********FOUND**********
         s = Str(a(m))
      Next m
      Print s
   End Sub
   Private Sub mnu_Eixt_Click()
      End
   End Sub
```

二、程序填空

第一题

【实验要求】

设计一个两位整数四则运算的检测程序，运行界面如图 8-4 所示。系统随机生成两位正整数，通过菜单选择加、减、乘、除四则运算，检测者通过文本框输入运算结果，用消息框提示运算结果是否正确。单击"下一题"菜单项更换运算数继续检测，单击"答案"菜单项系统给出正确结果，单击"退出"菜单项结束检测。

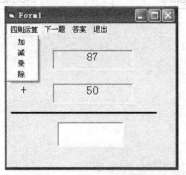 

图 8-4 四则运算检测运行界面

【实验步骤】

1. 界面设计

用菜单编辑器设计菜单，主要属性设置如表 8-2 所示。

表 8-2　　　　　　　　　　　　对象属性设置

菜单项对象	属　性	属　性　值
四则运算	名称	calc
…加	名称	add
…减	名称	miu
…乘	名称	mul
…除	名称	div

续表

菜单项对象	属 性	属 性 值
下一题	名称	next
答案	名称	ans
退出	名称	exit

在窗体上放 3 个标签（Label1~Label3），1 个线形控件（Line1）和 1 个文本框（Text1），其主要属性设置如表 8-3 所示。

表 8-3　　　　　　　　　　　　对象属性设置

对象	属性	属性值
Label1 和 Label2	Caption	空
	Alignment	2-Center
	BorderStyle	1-Fixed Single
Label3	Caption	+
Line1	BorderWidth	3
Text1	Text	空
	Aligment	2-Center

2. 编写事件代码

```
Dim a%, b%, c%, result
Private Sub add_Click()
   Label3.Caption = "+"
End Sub


Private Sub ans_Click()
   MsgBox result
End Sub


Private Sub div_Click()
   Label3.Caption = "÷"
End Sub


Private Sub miu_Click()
   Label3.Caption = "-"
End Sub


Private Sub mul_Click()
   Label3.Caption = "×"
End Sub

'**********SPACE**********
Private Sub 【?】()
   a = Int(Rnd * 100)
   b = Int(Rnd * 100)
   Label1.Caption = a
   Label2.Caption = b
   Text1.Text = ""
End Sub


Private Sub exit_Click()
```

```
      End
   End Sub

Private Sub Form_Load()
   Randomize
   a = Int(Rnd * 100)
   b = Int(Rnd * 100)
   Label1.Caption = a
   Label2.Caption = b
End Sub

Private Sub Text1_KeyPress(KeyAscii As Integer)
   Dim op As String
   op = Label3.Caption
   If KeyAscii = 13 Then
'**********SPACE**********
      Select Case 【?】
         Case "+"
            result = a + b
         Case "-"
            result = a - b
         Case "×"
            result = a * b
         Case "÷"
            result = a / b
      End Select
      c = Val(Text1.Text)
'**********SPACE**********
      If c = 【?】 Then
         MsgBox "恭喜你,答对了!", vbInformation, "提示"
      Else
         MsgBox "再想想,加油!", vbExclamation, "提示"
      End If
   End If
End Sub
```

第二题

【实验要求】

设计如图 8-5 所示的菜单,分别用 3 种不同的排序方法完成数组 a 中的 10 个整数按升序排列。

【实验步骤】

1. 界面设计

用菜单编辑器设计菜单,主要属性设置如表 8-4 所示。

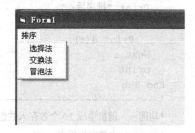

图 8-5 排序菜单运行界面

表 8-4　　　　　　　　　　　　　　对象属性设置

菜单项对象	属　　性	属　性　值
排序	名称	sort
…选择法	名称	Select
…交换法	名称	swap
…冒泡法	名称	bubble

2. 编写事件代码

```
Option Base 1
'----------------------------------------------------------
'功能：以下程序段利用随机函数生成 10 个 10~100 的整数，
'      然后用选择法将其从小到大排序。
'----------------------------------------------------------
Private Sub Select_Click()
  cls
  Const n = 10
  Dim a(1 To n) As Integer
  Dim i As Integer, j As Integer, t As Integer, min As Integer
  Randomize
  For i = 1 To n
'**********SPACE**********
    a(i) =【?】
  Next i
  Print "排序前: "
  For i = 1 To n
    Print a(i);
  Next i
  Print
  For i = 1 To n - 1
'**********SPACE**********
   【?】
    For j = i + 1 To n
      If a(j) < a(t) Then t = j
    Next j
'**********SPACE**********
    If 【?】 Then
      min = a(i): a(i) = a(t): a(t) = min
    End If
  Next i
  Print "排序后: "
  For i = 1 To n
    Print a(i);
  Next i
  print
End Sub
'----------------------------------------------------------
'功能： 随机生成 10 个数存入数组中，用交换法将数组从小到
'      大排序后输出
'----------------------------------------------------------
Private Sub Swap_Click()
  cls
  Dim a(10) As Integer
  Dim i, j, t As Integer
  Print "排序前: "
  For i = 1 To 10
    '**********SPACE**********
    a(i) = CInt(10 【?】 Rnd + 1)
    Print a(i);
  Next
```

```
        Print
        For i = 1 To 9
            '**********SPACE**********
            For j = 【?】 To 10
                '**********SPACE**********
                If a(i) > 【?】 Then
                    t = a(i): a(i) = a(j): a(j) = t
                End If
            Next j
        Next i
        Print "排序后："
        For i = 1 To 10
            Print a(i);
        Next
        Print
End Sub
'----------------------------------------------------------
'功能：下面的程序用"冒泡"法将数组 a 中的 10 个整数按从小到大
'       排列，请将程序补充完整。
'----------------------------------------------------------
Private Sub bubble_Click()
    cls
    Dim a
    a = Array(-2, 5, 24, 58, 43, -10, 87, 75, 27, 83)
    Print "排序前："
    For i = 1 To 10
        Print a(i);
    Next i
    Print
    For i = 10 To 2 Step -1
        '**********SPACE**********
        For 【?】
        '**********SPACE**********
            If 【?】 Then
                a1 = a(j + 1)
                a(j + 1) = a(j)
                a(j) = a1
            End If
        Next j
        '**********SPACE**********
        【?】
    Print "排序后："
    For i = 1 To 10
        Print a(i);
    Next i
    Print
End Sub
```

三、编程题

【实验要求】

设计一下拉式菜单和一弹出式（快捷）菜单，下拉式菜单用于设置文本框中文本的字体和字形，弹出式菜单用于控制文本框中文本的字号，运行界面如图 8-6 和图 8-7 所示。

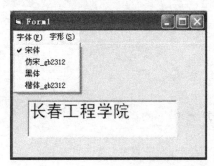 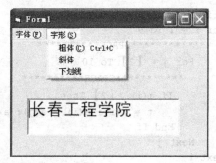

图 8-6 下拉式菜单运行界面

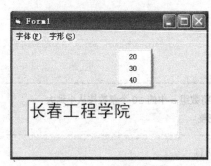

图 8-7 弹出式菜单运行界面

【实验步骤】

1. 界面设计

用菜单编辑器设计菜单,主要属性设置如表 8-5 所示。

表 8-5　　　　　　　　　　　　　　对象属性设置

菜单项对象	名称	索引	可见性	快捷键
字体(&F)	Font		可见	
…宋体	zt	1	可见	
…仿宋_gb2312	zt	2	可见	
…黑体	zt	3	可见	
…楷体_gb2312	zt	4	可见	
字形(&S)	zx		可见	
…粗体(&C)	ct		可见	Ctrl+C
…斜体	xt		可见	
…下划线	xhx		可见	
字号	h		不可见	
…20	zh	1	可见	
…30	zh	2	可见	
…40	zh	3	可见	

2. 编写事件代码

```
Private Sub ct_Click()
    If ct.Checked Then
```

```
      ct.Checked = False
      Text1.FontBold = False
   Else
      ct.Checked = True
      Text1.FontBold = True
   End If
End Sub
Private Sub Form_MouseDown(Button As Integer, Shift As Integer, x As Single, y As Single)
   If Button = 2 Then PopupMenu H
End Sub

Private Sub xt_Click()
   If xt.Checked Then
      xt.Checked = False
      Text1.FontItalic = False
   Else
      xt.Checked = True
      Text1.FontItalic = True
   End If
End Sub

Private Sub zh_Click(Index As Integer)
   Text1.FontSize = zh(Index).Caption
End Sub

Private Sub zt_Click(Index As Integer)
   Text1.FontName = zt(Index).Caption
   For i = 1 To 4
      If i = Index Then
         zt(i).Checked = True
      Else
         zt(i).Checked = False
      End If
   Next i
End Sub
Private Sub xhx_Click()
   If xhx.Checked Then
      xhx.Checked = False
      Text1.FontUnderLine = False
   Else
      xhx.Checked = True
      Text1.FontUnderLine = True
   End If
End Sub
```

8.3 测试题

一、判断题

1. 菜单中每一菜单项都是控件，可以通过单击菜单项或通过光标选择并按回车键，以触发 Click 事件。（   ）

2. "菜单编辑器"中至少要填"名称"和"标题"这两个框，才能正确完成菜单栏的设计。

3. 定义菜单项时，可以不设置分隔线的菜单项控件名称。（   ）
4. 如果一个菜单项的 Visible 属性为 False，则它的子菜单也不会显示。（   ）
5. 设计菜单中每一个菜单项分别是一个控件，每个控件都有自己的名字。（   ）

二、填空题
1. 如果建立菜单时在标题文本框中输入一个_____，那么显示时形成一个分隔符。
2. 在 VB 的菜单设计中，可以建立_____菜单和_____菜单。
3. 如果把菜单项的_____属性设置为 True，则该菜单项成为一个选定项。
4. 菜单的访问键（热键）指使用_____键和菜单项标题中的一个字符来打开菜单。建立热键的方法是，在菜单标题的某个字符前加上一个_____符号，则菜单中这一字符自动加上下划线表示该字符是热键字符。
5. 菜单控件（对象）只有一个_____事件。
6. 当菜单项的_____属性设为 False 时，则程序执行时该菜单不可见。
7. 显示弹出式菜单需用使用_____方法。

三、选择题
1. 设计下拉式菜单时，顶级菜单的可见性通常设置成（   ）。
   [A] 可见   [B] 不可见
   [C] 与弹出式菜单相同   [D] 没有可见性设置这项
2. 假定有一个菜单，名为 MenuItem，为了运行时使该菜单项失效，应使用的语句为（   ）。
   [A] MenuItem.Enabled = True   [B] MenuItem.Enabled = False
   [C] MenuItem.Visible = True   [D] MenuItem.Visible = False
3. 下列选项中不正确的是（   ）。
   [A] 每个菜单项都是一个对象，所以也有属性和事件
   [B] 菜单项的属性可以在属性窗口中设计
   [C] 每个菜单项都只有一个 Click 事件
   [D] 菜单编辑器中，标题可不输入，而名称必须输入
4. 在下列关于菜单的说法中，错误的是（   ）。
   [A] 每个菜单项是一个控件，与其他控件一样也有自己的属性和事件
   [B] 除了 Click 事件之外，菜单项还能响应其他的事件，如 DblClick 事件
   [C] 在程序执行时，如果菜单项的 Visible 属性为 False，则该菜单项不可见
   [D] 在程序执行时，如果菜单项的 Enabled 属性为 False，则该菜单项变成灰色，不能被用户选择
5. 在用菜单编辑器设计菜单时，必须输入的项有（   ）。
   [A] 快捷键   [B] 标题   [C] 索引   [D] 名称
6. 用户可以通过设置菜单项的（   ）属性值为 FALSE 来使该菜单项不可见。
   [A] Hide   [B] Checked   [C] Visible   [D] Enabled
7. 一个菜单项是不是一个分隔条，由（   ）属性决定。
   [A] Name（名称） [B] Caption   [C] Enabled   [D] Visible

第 9 章 文件操作

9.1 知识点

9.1.1 文件分类

(1) 顺序文件: 数据是以字符的形式存储,数据的写入、存放与读出的顺序是一致的。

(2) 随机文件: 随机文件由记录组成,每条记录等长,各数据项长度固定,每个记录有唯一的记录号,读/写文件时按记录号对记录读/写。

(3) 二进制文件。

9.1.2 顺序文件

1. 打开文件

Open 文件名 For [Append | Input | Output] As [#文件号]

Append: 为追加数据。Input: 为读取数据。Output: 为改写数据。

2. 关闭文件

Close [[#]文件号]

3. 写文件

Write #文件号, [输出项列表]

4. 读文件

Input #文件号, [输出项列表]

5. 常用函数

EOF、LOF、LOC 等。

EOF (文件号): EOF 函数值为 True, 到文件尾; EOF 函数值为 False, 没到文件尾。

LOF (文件号): 返回文件长度, 如果返回 0 则表示该文件为空。

LOC (文件号): 返回一个长整型数, 在已打开的文件中指当前读写位置。

9.1.3 随机文件

1. 打开文件

Open 文件名 For Random As #文件号[Len=记录长度]

2. 关闭文件

Close [[#]文件号]

3. 写文件

Put　#文件号，[记录号]，变量列表

4. 读文件

Get　#文件号，[记录号]，变量列表

5. 常用函数

FileLen、LOF、EOF、LOC。

FileLen（<文件名>）：文件长度测试函数。

EOF（文件号）：EOF 函数值为 True，到文件尾；EOF 函数值为 False，没到文件尾。

LOF（文件号）：返回文件长度，如果返回 0 则表示该文件为空。

LOC（文件号）：返回一个长整型数，在已打开的文件中指当前读写位置。

9.2　实验内容

文件实验

【实验目的】

1. 掌握顺序文件的打开和关闭。
2. 掌握顺序文件的读、写操作和编程技巧。
3. 掌握随机文件的打开和关闭。
4. 掌握随机文件的读、写操作和编程技巧。

一、程序改错

第一题

【实验要求】

使用顺序文件读写方式编写一个简单的记事本应用程序，在本程序文件夹中有一个名为 exam.txt 的文本文件。当单击"打开"按钮（Command1）时，程序将 exam.Txt 文件中的内容显示在文本框（Text1）中；当单击"新建"按钮（Command2）时，清空 Text1 中的内容，用户可以在 Text1 中进行编辑操作；当单击"保存"按钮（Command3）时，将 Text1 中的内容保存在 exam.txt 文件中；当单击"退出"按钮（Command4）时关闭窗体。运行界面如图 9-1 所示。改正所给程序段中的错误。

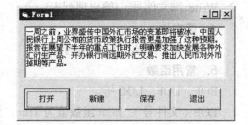

图 9-1　简单记事本运行界面

【实验步骤】

1. 界面设计

在窗体上添加一个文本框控件 text1 和 4 个命令按钮 Command1、Command2、Command3、Command4，其大小及位置关系如图 9-1 所示。控件属性设置如表 9-1 所示。

表 9-1   对象属性设置

对象	属性	属性值
Text1	Text	空
Command1、Command2、Command3、Command4	Caption	打开、新建、保存、退出
Command1、Command2、Command3、Command4	TabIndex	0、1、2、3

2. 编写事件代码

```
Option Explicit
Private Sub Command1_Click()    '打开按钮
    Dim A As String
    Text1 = ""
    Open App.Path & "\exam.txt" For Input As #1
    Do While Not EOF(1)
    '**********FOUND**********
       Input #0, A
       Text1 = Text1 + A
    Loop
    '**********FOUND**********
    Close #0
End Sub

Private Sub Command2_Click()    '新建按钮
    Text1 = ""
End Sub

Private Sub Command3_Click()    '保存按钮
    Open App.Path & "\exam.Txt" For Output As #1
    '**********FOUND**********
    Input #1, Text1
    Close #1
End Sub

Private Sub Command4_Click()    '退出按钮
    Unload Me
End Sub
```

第二题

【实验要求】

下面的程序段实现以下功能：单击窗体空白处建立一顺序文件，存放 10 名同学的学号和三门功课成绩，打开该文件，读取所有记录显示在窗口中，并计算总分和平均分。运行界面如图 9-2 所示，改正所给程序段中的错误。

图 9-2 运行界面

【实验步骤】

编写事件代码

```
Option Explicit
Private Sub Form_Click()
    Dim no%, c1%, c2%, c3%, i As Integer
    '**********FOUND**********
    Open "c:\2.txt" For Input As #1
    For i = 1 To 10
        no = InputBox("请输入学号")
        c1 = Val(InputBox("请输入数学成绩"))
        c2 = Val(InputBox("请输入语文成绩"))
        c3 = Val(InputBox("请输入外语"))
        Write #1, no, c1, c2, c3
    Next i
    Close #1
    '**********FOUND**********
    Open "c:\2.txt" For Output As #1
    For i = 1 To 10
        '**********FOUND**********
        Print #1, no, c1, c2, c3
        Print no, c1, c2, c3, c1 + c2 + c3, Format((c1 + c2 + c3) / 3, "#####.00")
    Next i
    Close #1
End Sub
```

二、程序填空

第一题

【实验要求】

窗体运行界面如图 9-3 所示，窗体中有两个 List 列表及 4 个命令按钮。程序功能：单击"产生随机数"按钮，产生 20 个随机数填入 List1 中；单击"保存"按钮，将 list1 中数据写到顺序文件中；单击"读出"按钮，将文件内容读到数组中并填入 List2 中；单击"结束"按钮，结束程序运行，根据程序功能将所给程序段的【?】处填写完整并运行该程序。

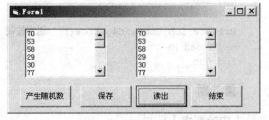

图 9-3 运行结果界面

【实验步骤】

1. 界面设计

在窗体上添加 2 个 List 控件、4 个命令按钮，其大小及位置关系如图 9-3 所示。属性设置如表 9-2 所示。

表 9-2　　　　　　　　　　　　　　对象属性设置

对象	属性	属性值
List1、List2	Text	空
Command1、Command2、Command3、Command4	Caption	产生随机数、保存、读出、结束
Command1、Command2、Command3、Command4	TabIndex	1、2、3、4

2. 编写事件代码

```
Dim d(1 To 20) As Integer
Private Sub Command1_Click()            ' "产生随机数"按钮
    Dim i As Integer
    List1.Clear
    For i = 1 To 20
    '**********SPACE**********
        d(i) = Int(1 + 99 *【?】)
        List1.AddItem d(i)
    Next i
End Sub
Private Sub Command2_Click()            ' "保存"按钮
    Dim i As Integer
    Open App.Path & "\MyFile3.txt" For Output As #1
    For i = 1 To 20
    '**********SPACE**********
        Write【?】, d(i)
    Next i
    Close #1
End Sub
Private Sub Command3_Click()            ' "读出"按钮
    Dim x As String
    Open App.Path & "\MyFile3.txt" For Input As #1
    List2.Clear
    '**********SPACE**********
    Do While Not【?】
        Input #1, x
        List2.AddItem x
    Loop
    Close #1
End Sub
Private Sub Command4_Click()            ' "结束"按钮
    End
End Sub
```

第二题

【实验要求】

设计如图 9-4 所示的窗体，利用随机文件编写一简单通讯录程序，实现如下功能。

（1）单击"打开"按钮，打开程序文件夹下的 Tel.dat 随机文件，若没有 Tel.dat 文件则建立该文件，若该文件有记录信息，取出第 1 个人的信息并显示在文本框中，正确打开文件后，该按钮设置成无效状态。

（2）单击"〉〉"和"〈〈"按钮可以向后或向前查看通讯录中的信息。

（3）单击"新建"按钮，文本框设置为空，文件指针指向文件尾。光标在姓名文本框等待输入。

（4）单击"保存"按钮将当前文本框信息添加到 Tel.dat 数据文件中。

图 9-4　简单通讯录运行界面

（5）单击"关闭"按钮，关闭数据文件，程序退出。

【实验步骤】

1. 界面设计

在窗体中添加 3 个标签 3 个文本框控件，6 个命令按钮 Command1、Command2…Command6。其大小及位置关系如图 9-4 所示。属性设置如表 9-3 所示。

表 9-3   对象属性设置

对象	属性	属性值
Text1、Text2、Text3	Text	空
Text1、Text2、Text3	MaxLength	4、11、12
Command1、Command2、Command3、Command4 Command5、Command6	Caption	打开、<<、>>、新建、保存、关闭
Command1、Command2、Command3、Command4 Command5、Command6	Enabled	True、False、False、False False、False
Command1、Command3、Command4、Text1 Text2、Text3 Command5、Command2 Command6	TabIndex	0、1、2 3、4 5、6 7、8

2. 编写事件代码

```
Private Type Tel
    Name As String * 8
    Number As String * 11
    QQ As String * 12
End Type

Private Sub Command1_Click()    '打开按钮事件
    Dim person As Tel
    Dim i As Integer
    '**********SPACE**********
    Open App.Path & "\student.dat" For 【?】As #1 Len = Len(person)

    '根据文件是否有记录设置各命令按钮有效状态
    If LOF(1) / Len(person) > 1 Then
        Call Command3_Click
    Else
        Call Command4_Click
        Command4.Enabled = False
        Command3.Enabled = False
        Command2.Enabled = False
    End If
    Command1.Enabled = False
    Command5.Enabled = True
    Command6.Enabled = True
End Sub

Private Sub Command2_Click()'  向前查看<<按钮事件
```

```
    Dim person  As Tel
    '**********SPACE**********
    Get #1, 【?】, person
    Text1.Text = person.Name
    Text2.Text = person.Number
    Text3.Text = person.QQ
    Call anniu
    Command4.Enabled = True
End Sub

Private Sub Command3_Click()    '向后查看>>按钮事件
    Dim person  As Tel
    Get #1, , person
    Text1.Text = person.Name
    Text2.Text = person.Number
    Text3.Text = person.QQ
    Call anniu
    Command4.Enabled = True
End Sub

Private Sub Command4_Click()'新建按钮事件
    Dim person  As Tel
    Text1.Text = ""
    Text2.Text = ""
    Text3.Text = ""
    '设置指针指向文件结尾
If Not EOF(1) And LOF(1) / Len(person) <> 0 Then
        Get #1, LOF(1) / Len(person), person
    End If
    Get #1, , person
    '设置按钮 >>和新建按钮不可用
    Command3.Enabled = False
    Command4.Enabled = False
    Text1.SetFocus
End Sub

Private Sub Command5_Click()  '保存按钮事件
    Dim person  As Tel
    Dim i%
    If Text1.Text <> "" And Text2.Text <> "" Then
      person.Name = Text1.Text
      person.Number = Text2.Text
      person.QQ = Text3.Text
      '**********SPACE**********
      Put 【?】, Loc(1), person
      If EOF(1) Then
         Call Command4_Click
      End If
    End If
    Call anniu
End Sub

Private Sub Command6_Click()  '关闭按钮事件
    '**********SPACE**********
```

```
        【?】
         End
    End Sub
    Private Sub anniu()
        '根据文件读取指针位置,设置<<和>>按钮是否有效
        Dim person As Tel
        If (Loc(1) < 2) Then
            Command2.Enabled = False
        Else
            Command2.Enabled = True
        End If
        If (Loc(1) = LOF(1) / Len(person)) Then
            Command3.Enabled = False
        Else
            Command3.Enabled = True
        End If
    End Sub
```

三、编程题

【实验要求】

在第 3 章编程实验的基础上,设计如图 9-5 所示的教师基本信息输入窗口,"核对"按钮与"清空"按钮功能不变,单击"保存"按钮,先将文本框中的教师基本信息保存到变量中,然后添加到右侧的列表框,最后存储到数据文件 teacher.dat 中(teacher.dat 文件可以是顺序文件,也可以选择随机文件)。窗体打开时,teacher.dat 中所有教师信息自动载入列表框中。根据功能编写下面程序段。

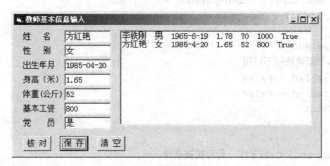

图 9-5 程序运行界面

【实验步骤】

1. 界面设计

在原实验基础上窗口右侧添加 1 个列表框 List1,List 属性设置为空。

2. 编写事件代码

```
'Private Sub Command2_Click()   '保存按钮事件
    '声明变量,变量名用英文单词或汉语拼音头写字符命名,注意选择变量数据类型。
    '********** Program *********


    '********** End *************
    '教师信息暂存变量中
    '********** Program *********
```

```
'********** End *************
'教师信息添加到列表框
'********** Program *********



'********** End *************
'教师信息存储在 teacher.dat 顺序文件中
'********** Program *********



'********** End *************
End Sub
Private Sub Form_Load()
    '打开 teacher.dat 文件,将所有教师信息加载到列表框后,关闭文件
    '********** Program *********



'********** End *************
End Sub
```

9.3 测试题

一、选择题

1. 以下能判断文件指针是否到达文件尾的函数是（   ）。
   [A] BOF　　　　　[B] LOC　　　　　[C] LOF　　　　　[D] EOF
2. 如果从数据文件 INPUT.DAT 中读取数据时,以下打开文件方式中,正确的是（   ）。
   [A] OPEN INPUT.DAT FOR INPUT AS #1
   [B] OPEN INPUT.DAT FOR OUTPUT AS #1
   [C] OPEN "INPUT.DAT" FOR INPUT AS #1
   [D] OPEN "I", #1 INPUT.DAT
3. 以下（   ）方式打开的文件只能读不能写。
   [A] Input　　　　[B] Output　　　　[C] Random　　　　[D] Append
4. 下列说明中,不属于随机文件特点的是（   ）。
   [A] 可以随意读取随机文件中任一记录的数据
   [B] 随机文件没有只读或只写的操作方式,随机文件只要一打开,就既可读又可写
   [C] 随机文件的操作是以记录为单位进行的
   [D] 随机文件的读、写操作语句与顺序文件的读写操作语句一样
5. 下列文件打开方式中，以顺序文件方式打开文件并执行写操作的语句是（   ）。
   [A] Open　　"c:\file1.dat"　　For　　Output　　as　　#1
   [B] Open　　"c:\file1.dat"　　For　　Input　　as　　#1
   [C] Open　　"c:\file1.dat"　　For　　Append　　as　　#1
   [D] Open　　"c:\file1.dat"　　For　　Write　　as　　#1

二、填空题

1. 随机文件以_____为单位读写数据；二进制文件以_____为单位读写数据。
2. EOF 函数判断_____是否到了文件结束标志.LOF 函数返回文件的_____。
3. 按照数据文件的存储方法的不同，文件可以分为3种不同的类型：_____、顺序文件和二进制文件。
4. 按文件号2打开顺序文件 SEQOLD.DAT，文件打开方式是读出数据的语句_____。

三、判断题

1. 文件按照数据编码方式可以分为 ASCII 码文件和二进制文件。（   ）
2. 若要新建一个磁盘上的顺序文件，可用 output，append 方式打开文件。（   ）
3. 随机文件中记录的长度是随机的。（   ）
4. 顺序文件中的记录一个接一个地顺序存放。（   ）
5. 执行打开文件的命令后，自动生成一个文件指针。（   ）
6. 顺序文件和随机文件的打开都使用 open 语句。（   ）
7. LOF 函数返回给文件分配的字节数。（   ）
8. 随机文件打开后，可以同时进行读/写操作。（   ）

四、程序填空题

1. 将 C 盘根目录下的一个文本文件"OLD.DAT"复制到新文件"NEW.DAT"中，并利用文件操作语句将"OLD.DAT"文件从磁盘上删除。

```
Private Sub Command1?_Click()
   Dim strl $
   '**********SPACE**********
   Open "c:\old.dat" 【?】As #1    '读入方式打开
   '**********SPACE**********
   Open "c:\new.dat" 【?】         '写方式打开
   '**********SPACE**********
   Do While 【?】                  '文件没有到结束标志
      '**********SPACE**********
      【?】                        '按行读入到字符串变量中
      Print #2,strl
   Loop
   '**********SPACE**********
   【?】
   '**********SPACE**********
   【?】                           '删除老文件
End Sub
```

2. 以下代码建立文件名为"c:\stud1.txt"的顺序文件，内容来自文本框，每按一次回车键写入一条记录，然后清除文本框的内容，直到文本框内输入"END"字符串。

```
Private Sub Form_Load()
   Open "c:stud1.txt" For Output As #1
   Text1= ""
End Sub

Private Sub Text1_KeyPress(KeyAscii As Integer)
   If KeyAscii=13 Then
```

```
'**********SPACE**********
    If 【?】Then        '大小写均可
        Close #1
        End
    Else
'**********SPACE**********
        【?】       '文本框内容写到文件中
        Text1= ""
    End If
End If
End Sub
```

3. 窗体中有两个命令按钮,单击 cmdput 命令按钮时,执行如下的事件过程,该过程的功能是用输入对话框输入 5 个学生的学号和姓名,存储到指定的随机文件中,将程序补充完整。

```
'**********SPACE**********
Private Type 【?】
    Number As String * 10
    Name As String * 10
End Type
Private Sub cmdput_click()
    Dim Title As String
    Dim Str1 As String
    Dim str2 As String
    Dim stu  As Student
    Dim i As Integer
    Open App.Path & "\student.dat" For Random As #1 Len = Len(stu)

    Title = "写记录到随机文件"
    Str1$ = "请输入学生号"
    str2$ = "请输入学生名"
    For i = 1 To 5
'**********SPACE**********
        【?】.Number = InputBox(Str1$, Title$)
        stu.Name = InputBox$(str2$, Title$)
        Put #1, i, stu
    Next i
'**********SPACE**********
    Close 【?】
End Sub
```

第 10 章 数据库应用程序设计

10.1 知识点

数据控件的使用

1. 使用 VB 可视化数据管理器，建立 Access 数据库及表。
2. 数据控件 Data 常用属性。Connect、DatabaseName、RecordSource、ReadOnly、Recordtype、BOFAction、EOFAction 属性。
3. 数据控件 Data 记录集方法。AddNew、Delete、Edit、Move、Movefirst、Movelast、Movenext、MovePrevious、Move、Refresh、Find、Update、Close 方法。
4. 常用数据绑定控件。标签、文本框、复选框、组合框、列表框等控件的 DataSource 属性和 DataFleld 属性及 DataGrid 控件用法。
5. ADO 数据库访问技术，数据控件 Adodc 记录集常用的方法、属性。
6. 常用 SQL 查询语句。

10.2 实验内容

10.2.1 用 Adodc 控件连接数据库和表实验

【实验目的】
1. 掌握 Adodc 数据控件连接数据库和表的方法。
2. 掌握 Adodc 数据记录集 Recordset 常用方法及属性。
3. 掌握标准控件和 Adodc 连接及 DataGrid 控件用法。

【实验要求】
在 VB 环境下通过 ADO 技术访问外部 Access 数据库。设计窗体如图 10-1 所示。用 Adodc 控件连接数据库和表。通过 Adodc 数据记录集 Recordset 的方法对"成绩表"进行操作。

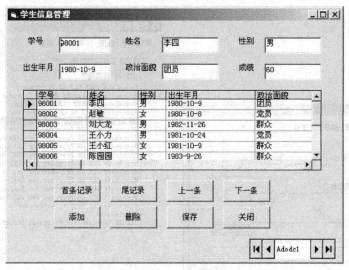

图 10-1　学生信息管理窗体

【实验步骤】

1. 通过 Microsoft Office Access 2003 建立"学生数据库",按表结构建立 student 表,并输入记录,表结构如表 10-1 所示。

表 10-1　　　　　　　　　　　　学生"成绩表"结构

字段	类型	宽度
学号	文本	8
姓名	文本	8
性别	文本	2
出生年月	日期时间	8
政治面貌	文本	10
成绩	数字	整型

2. 界面设计:在"工程"菜单中选"部件",选择"Microsoft ADO Data Controls 6.0",添加 Adodc 数据控件到工具箱,然后在窗体中添加数据控件 Adodc1。在"工程"菜单中选"部件",选"Microsoft DataGrid Control 6.0",将 DataGrid 控件添加到工具箱,然后在窗体中添加 DataGrid1。再添加 6 个标签、6 个文本框和 8 个命令按钮。

3. 主要属性设置:用数据控件对象 Adodc1 连接数据库,Adodc1 的属性设置如下。

(1) 右键单击 Adodc1 控件,选择 ADODC 属性,打开如图 10-2 所示的属性页,选择"通用"选项卡,选择"使用连接字符串"项,单击"生成"按钮。

(2) 在如图 10-3 所示的"数据链接属性"对话框中选"提供程序"选项卡,选择"Microsoft Jet 4.0 OLE DB Provider",单击"下一步"按钮。

(3) 在如图 10-4 所示的"连接"选项卡中,选择 "学生数据库",并进行"测试连接"。

(4) 在如图 10-5 所示的"记录源"选项卡中,记录源的命令类型选"2-adCmdTable",表或存储过程名称选"成绩表",单击"确定"按钮。

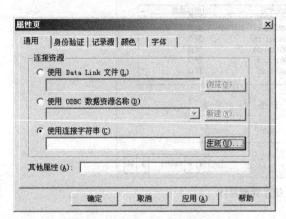

图 10-2　Adodc1 数据控件属性页

图 10-3　数据连接程序

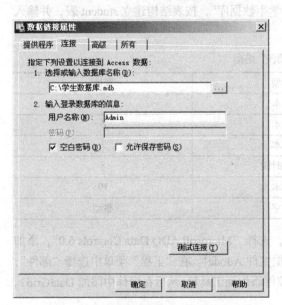

图 10-4　数据连接属性中连接数据库

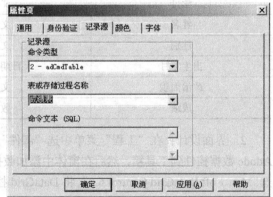

图 10-5　通过"属性页"中的"记录源"连接表

4. 其他控件的属性设置如表 10-2 所示。

表 10-2　　　　　　　　　　　窗体及控件属性表

对象	属性名	属性值
Text1～Text6	DataSource	Adodc1
	DataFleld	分别绑定 Adodc1 中的相应字段：学号、姓名、性别、出生年月、政治面貌、成绩
Command1～Command8	Caption	首条记录、尾记录、上一条、下一条、添加、删除、保存、关闭

5. 编写事件代码。

```
Private Sub Command1_Click()    ' "首条记录"命令按钮的单击事件过程
    Adodc1.Recordset.MoveFirst    ' 指向第一个记录
End Sub
Private Sub Command2_Click()    ' "尾记录"命令按钮的单击事件过程
    '**********SPACE**********
    Adodc1.Recordset.【?】  ' 指向尾记录
End Sub
Private Sub Command3_Click()    ' "上一条"命令按钮的单击事件过程
    If Adodc1.Recordset.BOF Then
        MsgBox "已到记录头"
    Else
        '**********SPACE**********
        Adodc1.【?】          ' 指向上一个记录
    End If
End Sub
Private Sub Command4_Click()    ' "下一条"命令按钮的单击事件过程
    '**********SPACE**********
    If 【?】 Then
        MsgBox "已到记录尾"
    Else
        Adodc1.Recordset.MoveNext  ' 指向下一条记录
    End If
End Sub
Private Sub Command5_Click()    ' "添加"命令按钮的单击事件过程
    '**********SPACE**********
    Adodc1.Recordset.【?】  ' 添加新记录
End Sub
Private Sub Command6_Click()    ' "删除"命令按钮的单击事件过程
    '**********SPACE**********
    【?】                ' 删除记录。
End Sub
Private Sub Command7_Click()    ' "保存"命令按钮的单击事件过程
    Adodc1.Recordset.Update     ' 更新记录
End Sub
Private Sub Command8_Click()    ' "关闭"命令按钮的单击事件过程
    '**********SPACE**********
    Adodc1.Recordset.【?】  ' 关闭记录集
    Unload Me
End Sub
```

10.2.2 设计学生信息查询窗体

【实验目的】

掌握在 VB 环境下，使用 SQL 查询语句实现数据库查询。

【实验要求】

在实验 10.2.1 数据环境基础上，用 SQL 查询语句实现查询功能。设计窗体如图 10-6 所示。

图 10-6　学生信息查询窗体

【实验步骤】

（1）主要界面设置：在"工程"菜单中选"部件"，选择"Microsoft ADO Data Controls 6.0"，添加 Adodc 数据控件到工具箱，然后在窗体中添加对象 Adodc1，在"工程"菜单中，选择"部件"中 Microsoft DataGrid Control 6.0，将添加到工具箱，然后在窗体中添加 DataGrid1。

（2）主要属性设置：设置和 10.2.1 相同的数据环境，用 Adodc1 数据控件连接"学生数据库"，测试连接成功后，选择"记录源"选项卡，命令类型选择"8_adcmdunknown"，在命令文本中填写 SQL 语句：select * from 成绩表，如图 10-7 所示。其他控件属性如表 10-3 所示。

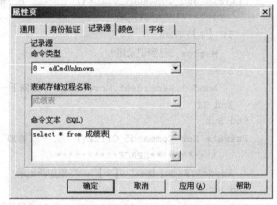

图 10-7　通过"属性页"中 SQL 查询语句连接表

表 10-3　　　　　　　　　　窗体及控件属性表

对象	属性名	属性值
Text1～Text6	DataSource	Adodc1
	DataFleld	分别绑定 Adodc1 中的相应字段：学号、姓名、性别、出生年月、政治面貌、成绩
DataGrid1	DataSource	Adodc1
Text_xm	Text	空格
Text_xb	Text	空格
Text_zz	Text	空格
Optxm	Caption	姓名
Optxb	Caption	性别
Optzz	Caption	政治面貌

3．编写事件代码

```
Private Sub Command1_Click()
```

```
        If Optxm.Value Then         ' 选择"姓名"单选按钮
            Adodc1.RecordSource = "select * from 成绩表 where 姓名='" + Trim(text_xm.Text) +
"' order by 学号"
            Adodc1.Refresh              ' 刷新查找记录
        ElseIf Optxb.Value Then      ' 选择"性别"单选按钮
        '按性别查找,并将查询结果按学号排序。
            '**********SPACE**********
            Adodc1.RecordSource = 【?】
            '**********SPACE**********
            Adodc1. 【?】              ' 刷新查找记录
            '**********SPACE**********
        ElseIf 【?】 Then            ' 选择"政治面貌"单选按钮
            Adodc1.RecordSource = "select * from 成绩表 where 政治面貌='" + Trim(Text_zz.Text)
+ "' order by 学号"
            '**********SPACE**********
            【?】                      ' 刷新查找记录
        End If
    End Sub
```

10.3 测试题

1. 在 Data 数据控件中,选择连接数据库类型的属性是(   )。

   [A] Connect   [B] DatabaseName   [C] RecordSource   [D] Recordtype

2. 通过数据记录集,修改表中当前记录的方法是(   )。

   [A] Edit   [B] Update   [C] Delete   [D] AddNew

3. 下面对 ADO 技术描述不正确的是(   )。

   [A] ADO 是 ActiveX Data Objects 的缩写,是一种数据库访问接口技术

   [B] 通过 ADO 技术只能访问 Access 数据库

   [C] 在 VB 环境下,通过 ADO 访问 Access 数据库,提供的数据链接程序为
       "MicrosoftJet4.0 OLE DB Provider"

   [D] 通过 ADO 技术可以执行 SQL 查询语句访问 Access 数据库

4. 从 Data1 记录集中查找第一个男生,正确的 SQL 查询语句是(   )。

   [A] Data1.Recordset.FindFirst 性别="男"

   [B] Data1.Recordset.FindLast 性别="男"

   [C] Data1.Recordset.FindNext 性别="男"

   [D] Data1.Recordset.FindPrevious 性别="男"

5. 用 Data1 数据控件连接表,判断记录指针是否指向表的顶部(首记录之前),应使用(   )。

   [A] Data1.Recordset.BOF=True   [B] Data1.Recordset.EOF= True

   [C] Data1.BOF=True   [D] Data1.EOF= True

6. 用 Data1 数据控件连接表,判断记录指针是否指向表的尾部(末记录之后),应使用(   )。

   [A] Data1.Recordset.BOF=True   [B] Data1.Recordset.EOF= True

   [C] Data1.BOF=True   [D] Data1.EOF= True

7. 在"职工表"中查询所有的男职工记录，正确的 SQL 查询语句是（　　）。
   [A] Select * from 职工表 where 性别="男"
   [B] Select * from 职工表 where 性别="男"
   [C] Select * from 职工表性别="男"
   [D] Select * where 性别="男"

8. 查询"职工表"中女党员的姓名和出生年月，正确的 SQL 查询语句为（　　）。
   [A] Select 姓名，出生年月 from 职工表 where"女党员"
   [B] Select 姓名，出生年月 from 职工表 where 性别="女"，政治面貌="党员"
   [C] Select 姓名，出生年月 from 职工表 where 性别="女"and 政治面貌="党员"
   [D] Select 姓名，出生年月 from 职工表 where 性别="女"or 政治面貌="党员"

9. 查询"成绩表"中所有的字段内容，并按成绩升序排列，正确的 SQL 查询语句为（　　）。
   [A] Select * from 成绩表 order by 成绩
   [B] Select * from 成绩表 order by 成绩 desc
   [C] Select * from 成绩表 by 成绩 asc
   [D] Select * from 成绩表 by 成绩 desc

10. 查询"成绩表"中的成绩最高分，正确的 SQL 查询语句为（　　）。
    [A] Select * from 成绩表 where 成绩=max(成绩)
    [B] Select * from 成绩表 where max(成绩)
    [C] Select max(成绩) from 成绩表 where 成绩=max(成绩)
    [D] Select max(成绩) as 最高分 from 成绩表

第 11 章
Visual Basic 综合性实验

【实验目的】
1. 加深对所学的 VB 程序设计方法和对象的属性、事件和方法的理解与掌握。
2. 提高综合运用所掌握的知识解决实际问题能力。

【实验要求】

以下题目作为综合实验的参考题目，并只给出基本要求，学生可根据所学 VB 知识自行完成实验题目的选定和运行界面及程序的设计。

设计过程中，应综合运用所学知识，将理论和实践有机的结合起来，提高用所学的 VB 知识解决实际问题的能力，为今后解决工作和日常生活中遇到的问题打下良好的基础。

综合实验参考题目：

1. 记事本的设计与实现

实现新建、打开、保存文件和打印功能，能够对文本进行字体设置，对文本实现查找功能。

2. 计算器的设计与实现

计算器要求实现加、减、乘、除、求乘方、正弦、余弦、正切、余切、对数等功能。

3. 猜数字游戏的设计与实现

产生 4 个随机数字，要求不能重复，用户可以猜 10 次，每次给出"位置不对数字对"和"位置数字都对"的数字个数，10 次未猜中给出正确数字，游戏可重新开始。

4. 指法游戏的设计与实现

指法游戏的设计要求随机生成一系列运动的字符，然后用户对照相应的字符输入，用户输入完成以后统计用户输入的速度和准确率，用户可以重新开始游戏或终止游戏。

5. 文件复制与删除的设计与实现

能够选择计算机中任意文件夹，将其包含的文件复制到任意其他文件夹（当文件夹不存在时可以创建）；能够删除任意选中文件夹中的所有文件。

6. 进制转换的设计与实现

实现不同进制（二、八、十、十六）的转换。要求可输入任意"变换前的数据"，并能指定其进制类型；选择不同的"变换后的数据"的进制，使"变换结果"中显示进制变换后的数据。

7. 小型考试系统的设计与实现

能够随机生成 10 道数学计算题（要求有加、减，每一道题的操作数为 2 个），对用户结果进行判断，并能最终计算总成绩，要求用户在规定的时间内做完试题（每题答题时间和总时间均有限制），超时自动终止并给出成绩。

8. 乒乓球游戏的设计与实现

用窗体做乒乓球案，用图形工具绘制乒乓球和球拍，一侧球拍可上下移动击球，球在两个球拍之间移动，当球拍未击中乒乓球时游戏结束，游戏进行过程中可随时暂停。

9. 屏幕保护程序的设计与实现

建立一屏幕保护程序，能够在计算机屏幕（图片）上随机绘制不同颜色的圆，随时间变化圆的数目逐渐增加，并能通过滚动条设置产生圆的大小。

还可实现"流星雨"屏保、"逐渐展开式"屏保、"孔雀开屏式"屏保等。

10. 图像浏览器的设计与实现

建立一图像浏览器，可以浏览 BMP 图片、JPG 图片、GIF 图片和 ICO 图标。对于大图片可以拖动滚动条进行浏览。如果同一个文件夹有多张图片，可以单击"上一张"和"下一张"按钮进行浏览。

11. 画图程序的设计与实现

建立一画图程序，通过打开对话框打开一 BMP 文件，能够绘制直线、矩形、圆等曲线；通过滚动条能够设置绘制曲线的颜色（调整 R、G、B 的值）。

12. 通讯簿的设计与实现

建立一通讯簿管理程序，将联系人的信息存入随机文件，能够实现联系人的添加、修改、删除和查找等功能。

还可建立学生管理程序、职工管理程序、客户信息管理程序、物资管理程序等。

第 12 章
Visual Basic 课程设计基础

12.1 概述

VB 课程设计是教学过程中的重要组成部分,用来培养学生运用所学的 VB 基本理论和基本技能,去分析问题和解决问题的能力。它与理论教学彼此配合,相辅相成,是教学过程的继续、深化和检验。它的实践性和综合性是其他环节所不能代替的。通过 VB 课程设计,使学生学会如何进行问题的调查研究、查阅收集资料;理论分析、设计;上机编程、调试和运行;报告的组织编写等。这不仅巩固加深 VB 理论知识,使理论和实践有机结合起来,而且也培养了学生应用 VB 开发设计应用程序的兴趣。提高了解决实际问题的能力,为以后的学习、工作打下坚实的基础。

12.2 课程设计的要求

一、课程设计的目的

每个学生在老师的指导下,通过 VB 课程设计,完成一定的设计任务。使学生对 VB 有一个综合全面的认识,主要锻炼以下 5 个方面的能力。

(1) 培养学生综合运用 VB 知识的能力。
(2) 培养学生进行应用系统分析设计和计算研究的能力。
(3) 提高查阅文献资料的能力。
(4) 提高学生编写、调试程序的能力。
(5) 提高学生进行设计总结和撰写报告的能力。

二、课程设计的一般步骤及要求

课程设计过程分为:前期准备阶段、选题和资料收集阶段、分析和计划阶段、设计阶段、编码调试阶段、报告撰写阶段、验收批阅阶段。具体内容和任务如下。

1. 准备工作阶段

根据 VB 课程设计大纲要求,在课程设计前,由指导教师编制 VB 课程设计任务书,并下达到每个学生手中。要求学生明确任务,进一步熟悉 VB 中的相关知识。

2. 资料收集阶段

学生接到课程设计任务书后,要全面具体地了解任务要求,确定设计题目,查询相关技术资料和文献，熟悉所需要的软硬件环境等相关知识。

3. 设计阶段

本阶段的工作主要包括：总体设计、详细设计。

总体设计阶段的任务是：在确定设计题目后，首先要明确系统的目标及实现方式，确定系统的功能，采用"自顶向下，逐步细化"的原则，将系统的功能模块化，画出系统的功能模块图，涉及到数据库时，要给出数据库的表结构及表间关系。

详细设计阶段的任务是：根据各个模块的功能给出具体解决方案和算法。要求方案设计合理，算法结构清晰简炼，界面友好，操作方便，并具有较好的可维护性。

4. 编码调试阶段

本阶段的任务主要包括：上机设计数据结构、VB 窗体界面设计，编写事件代码及系统的调试运行。

5. 课程设计报告编写阶段

课程设计报告是整个课程设计过程的总结性资料，撰写的质量直接影响到课程设计的好坏。课程设计报告要求层次分明，图表清晰，过程完整，阐述准确、文字通顺。

6. 课程设计成果的验收、及评阅阶段

课程设计完成后，由指导老师负责在计算机上验收学生设计的应用系统，包括设计界面、功能的实现、程序代码、问题的答辩等情况，最后评阅课程设计报告。

三、课程设计报告主要内容

（1）课程设计的目的及意义。
（2）课程设计的软硬件环境。
（3）总体设计思路及简要说明，描述数据库结构，给出系统功能模块图。
（4）详细设计包括：各模块功能的具体描述及实现方法、算法流程图、设计界面。
（5）系统的编程调试和运行。给出设计系统的文件名、程序清单及说明。
（6）课程设计的总结体会。
（7）教师评语及成绩。

12.3 课程设计预备知识

1. 熟悉 VB 程序设计基本语法和结构。
2. 熟悉 VB 标准控件和高级控件的用法（属性、事件和方法）。
3. 熟悉 Access 数据库及表的操作，熟悉 VB 中的 ADO 数据库访问技术及 SQL 查询语句。

12.4 课程设计参考题目

一、数据库管理类

这类题目的设计均和数据库有关，其基本功能包括数据的添加、修改、删除、浏览、查询、计算统计、打印报表等功能。常见的题目有：通讯录管理系统、人事档案管理系统、职工工资管理系统、图书管理系统、职工考勤管理系统、毕业生招聘管理系统、运动会管理系统、学生班级

管理系统、学生公寓管理系统、学生信息管理系统、学生成绩管理系统、学生选课系统、宾馆客房管理系统、超市管理系统、小型仓库管理系统、客户信息管理系统、物资管理系统、商品进销存管理系统、列车时刻表查询系统、小型考试系统等。

二、游戏类

例如，打字游戏、猜数字游戏、贪吃蛇、俄罗斯方块、快速配对游戏、拼图游戏等。

三、办公类

例如，简易文本编辑器、简易计算器、科学计算器、贷款计算器、图片浏览器、电子台历等。

第 13 章 Visual Basic 课程设计实例

VB 作为一种面向对象的可视化编程工具，越来越多地应用于数据库应用系统的开发。下面以"人事管理系统"为例，介绍 VB 环境下开发数据库应用系统的过程，包系统总体设计、数据库设计、详细设计、编码调试运行等过程。

13.1 系统总体设计

"人事管理系统"的开发总体任务是实现单位职工信息的系统化、规范化和自动化管理。

一、软硬件环境支持

硬件环境采用微机系统；软件环境采用 Windows xp、VB 6.0 和 Microsoft Office Access 2007。

二、系统主要功能包括

人员信息管理：包括人员档案基本信息的增加、删除、更新，不同方式的查询及报表的打印功能。

工资管理：包括工资调整、工资核算、工资的查询功能。

考勤管理：主要用于人员的请假、迟到管理。

三、系统功能模块图

系统功能模块图如图 13-1 所示。

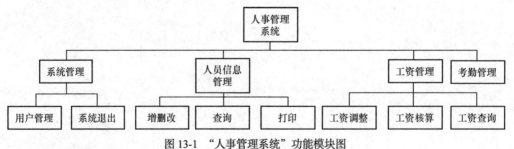

图 13-1 "人事管理系统"功能模块图

四、ADO 技术应用

在 VB 环境下访问数据库，其中应用广泛的是使用 ADO 技术访问 Access 数据库。它具有较强的功能、通用性好、效率高、占用内存空间小等特点。使用 ADO 访问数据库主要有两种方式，一种是使用 ADO 的 Data 控件，通过对控件的绑定来访问数据库中的数据，即非编程访问方式；另一种是使用 ADO 对象模型，通过定义对象和编写代码来实现对数据的访问，即编程访问方式。

13.2 数据库设计

数据库在信息管理系统中占有非常重要的地位。数据库结构设计的好坏直接影响到系统的效率和功能的实现。

下面以"人事管理数据库"为例,说明数据库的设计过程。首先分析用户的数据需求,将数据规范化,确立数据结构。然后建立数据的概念模型:E-R 模型,如图 13-2 所示。在此基础上建立数据的逻辑模型,其关系模式如下。

人员信息表(<u>编号</u>,姓名,性别,出生年月,政治面貌,职称,部门,文化程度,籍贯,家庭地址,联系电话),具体表结构如表 13-1 所示。

工资表(<u>编号</u>,基本工资,奖金,津贴,医疗保险,公积金,会员费,发放日期),具体表结构如表 13-2 所示。

考勤表(<u>编号</u>,请假天数,迟到次数,<u>月份</u>),具体表结构如表 13-3 所示。

用户表(<u>用户名</u>,密码),具体表结构如表 13-4 所示。

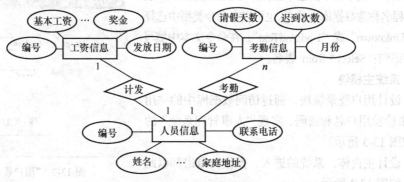

图 13-2　人事管理系统 E-R 图

表 13-1　"人员信息表"结构

字段名	类型	大小	说明
编号	文本	6	主键
姓名	文本	8	
性别	文本	2	
出生年月	日期时间	默认	
政治面貌	文本	10	
职称	文本	10	
部门	文本	10	
文化程度	文本	8	
籍贯	文本	10	
家庭地址	文本	50	
联系电话	文本	20	

表 13-2　"工资表"结构

字段名	类型	大小	说明
编号	文本	6	主键
基本工资	数字	单精度	
奖金	数字	单精度	
津贴	数字	单精度	
医疗保险	数字	单精度	
公积金	数字	单精度	
会员费	数字	单精度	
发放日期	日期时间	默认	

表 13-3　"考勤表"结构

字段名	类型	大小	说明
编号	文本	6	主键
请假天数	数字	整型	
迟到次数	数字	整型	
月份	数字	整型	主键

表 13-4　"用户表"结构

字段名	类型	大小	说明
用户名	文本	10	主键
密码	文本	10	

13.3　详细设计

一、创建表

通过 Microsoft Office Access 2007 建立 Access 数据库及表。

二、通过 ADO 技术访问外部 Access 数据库

通过"工程"菜单中"部件"命令项,选择"Microsoft ADO Data Controls 6.0",在窗体中添加数据控件 Adodc,用快捷键打开其属性页,在"通用"选项卡中选择"连接字符串",单击"生成",选择"Microsoft Jet 4.0 OLE DB Provider",单击"下一步"按钮,选择数据库名称,测试连接数据库成功后。选择"记录源"选项卡,其中命令类型中选择"2-adcmdTable",表或存储过程名称选数据库中的表。也可以在命令类型中选择"8-adcmdUnknown"或"1-adcmdText",在命令文本中填写 SQL 查询语句:Select * from 表名。

三、系统主模块

(1)设计用户登录模块,通过访问数据库中的"用户表",来检验用户名和密码,实现"人事管理系统"的登录,如图 13-3 所示。

(2)设计主窗体,系统的进入、操作和退出都以此为界面,如图 13-4 所示。

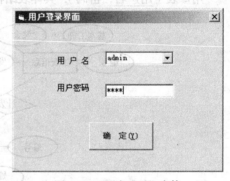

图 13-3　"用户登录"窗体

图 13-4　"人事管理系统"主窗体

（3）在主窗体中设计主菜单，实现各模块的调用，菜单结构如表 13-5 所示。

表 13-5　　　　　　　　　　　　　　菜单设计说明

菜单项名称结构	菜单项标题	说明
Xtgl	系统管理	一级菜单项
Xt(0)	用户管理	二级菜单项
Xt(1)	系统退出	二级菜单项
Rygl	人员信息管理	一级菜单项
Ry(0)	人员调整	二级菜单项
Ry(1)	人员查询	二级菜单项
Ry(2)	信息打印	二级菜单项
Gzgl	工资管理	一级菜单项
Gz(0)	人员工资调整	二级菜单项
Gz(1)	人员工资核算	二级菜单项
Gz(2)	工资查询	二级菜单项
Kqgl	考勤管理	一级菜单项
Kq(0)	人员考勤	二级菜单项

四、系统管理模块

访问数据库中的"用户表"，实现用户名和密码的修改功能，如图 13-5 所示。

五、人员信息管理模块

人员信息管理模块包括人员信息编辑子模块、人员信息查询子模块、报表打印子模块。

（1）人员信息的编辑模块：访问数据库中的"人员信息表"，实现记录的增加、删除、修改功能，如图 13-6 所示。

图 13-5　"用户管理"窗体

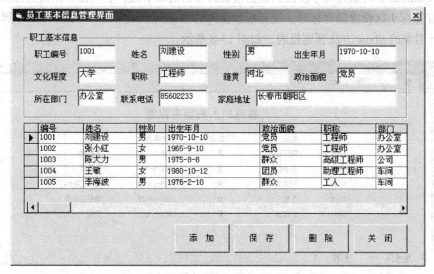

图 13-6　"人员信息编辑"窗体

（2）人员信息查询模块：访问数据库中的"人员信息表"，实现数据的查询浏览功能，可以采用姓名、性别、出生年月、部门、文化程度、职称等不同条件进行查询。照片的处理是将照片图像的主文件名（扩展名为 JPG）和表中编号相对应，用图片框的 LoadPicture()调用，如图 13-7 所示。

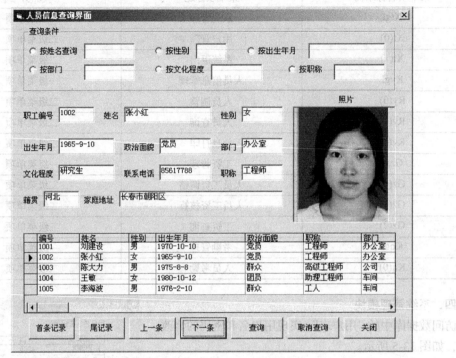

图 13-7 "人员信息查询"窗体

（3）报表打印模块：利用报表设计器来制作报表，从"工程"中选择"添加 Data Report"，将报表设计器加入到当前工程中。报表中的数据来源需要利用数据环境设计器与数据库的连接。步骤是：从"工程"菜单中选择 "添加 Data Enviroment"，在连接中选择"人事管理数据库"文件，完成与数据库的连接，然后用产生的 Command 对象连接"人员信息表"。再将数据环境设计器中 Command 对象内的字段拖到数据报表设计器的细节区，利用标签在报表标头区插入报表名称，在页标头区设置顶部标题。设置报表窗体的 DataSource 属性和 DataMember 属性。利用 DataReport 对象的 Show 方法预览报表，如图 13-8 所示。

图 13-8 人员信息打印报表

六、工资管理模块

工资管理模块包括调整工资子模块、工资核算子模块、工资查询子模块。

（1）调整工资模块：通过访问数据库中的"工资表"，实现对基本工资、奖金、津贴、医疗保险、公积金、会费的变动修改功能，如图13-9所示。

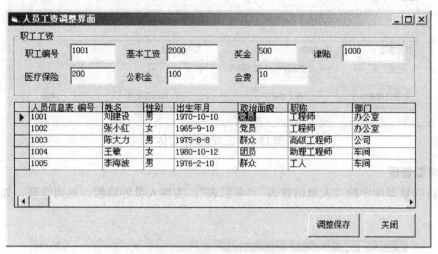

图 13-9 "工资调整"窗体

（2）工资核算模块：通过访问数据库中的"人员信息表""工资表"和"考勤表"，根据考勤情况，核算实发工资，如图13-10所示。

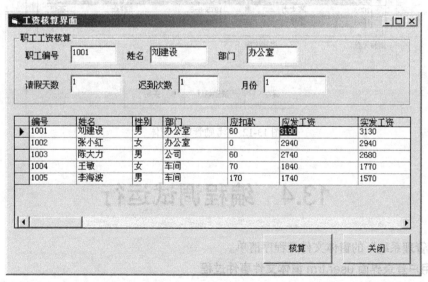

图 13-10 "工资核算"窗体

（3）工资查询模块：通过访问数据库中的"人员信息表""工资表"。实现基本工资的最大值、最小值、总和、平均值及应发工资的计算功能；实现按基本工资的升序、降序功能；实现按姓名、性别、部门、职称等字段的查询功能，如图13-11所示。

图 13-11 "工资查询"窗体

七、考勤管理

通过访问数据库中的"人员信息表""考勤表",实现人员的请假、迟到管理,如图 13-12 所示。

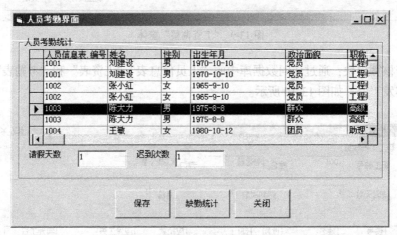

图 13-12 "考勤管理"窗体

13.4 编程调试运行

"人事管理系统"的窗体文件及程序清单。

一、用户登录界面 user.frm 窗体文件事件过程

```
Private Sub Passcmd_Click()     '用户登录
  Adodc1.RecordSource = "select * from 用户表"
  Adodc1.Refresh
  myname = Adodc1.Recordset.Fields("用户名")
  mypassword = Adodc1.Recordset.Fields("密码")
  If Combo1.Text = myname Then
    If UCase(Text1.Text) = mypassword Then
```

```
            MsgBox "口令正确！欢迎使用本系统"
            Load main
            main.Show
            Unload user
         Else
            Beep
            MsgBox "口令错误，请重新登录！"
            End
         End If
      Else
         Beep
         MsgBox "用户名错误，请重新登录！"
         End
      End If
End Sub
```

二、用户管理界面 yhgl.frm 窗体文件事件过程

```
Private Sub Userclosecmd_Click()     ' 关闭窗口
   Unload Me
End Sub
Private Sub Usereditcmd_Click()      ' 修改用户信息
   Adodc1.Recordset.Update
End Sub
```

三、系统主界面及菜单调用 main.frm 窗体文件事件过程

```
Private Sub xt_Click(Index As Integer)    ' 系统管理
   Select Case Index
      Case 0   ' 用户管理
         Load yhgl
         yhgl.Show
      Case 1   ' 系统退出
         End
   End Select
End Sub
Private Sub ry_Click(Index As Integer)    ' 人员管理
   Select Case Index
      Case 0   ' 人员信息编辑
         Load rybj
         rybj.Show
      Case 1   ' 人员信息查询
         Load rycx
         rycx.Show
      Case 2   ' 人员信息打印
         DataReport1.Show
   End Select
End Sub
Private Sub gz_Click(Index As Integer)   ' 工资管理
   Select Case Index
      Case 0   ' 工资调整。
         Load gztz
         gztz.Show
      Case 1   ' 工资核算。
```

```
            Load gzhs
            gzhs.Show
        Case 2    ' 工资查询。
            Load gzcx
            gzcx.Show
        End Select
End Sub
Private Sub kq_Click(Index As Integer)    ' 考勤管理
    Load rykq
    rykq.Show
End Sub
```

四、人员编辑界面 rybj.frm 窗体文件事件过程

```
Private Sub Addcmd_Click()          ' 人员添加
    Adodc1.Recordset.AddNew
End Sub
Private Sub Editcmd_Click()         ' 人员修改
    Adodc1.Recordset.Update
End Sub
Private Sub Delecmd_Click()         ' 人员删除
    Adodc1.Recordset.Delete
End Sub
Private Sub Editclosecmd_Click()    ' 关闭窗口
    Unload Me
End Sub
```

五、人员查询界面 rycx.frm 窗体文件事件过程

```
Private Sub Topamd_Click()          ' 移动到首记录
    Adodc1.Recordset.MoveFirst
    zp = Adodc1.Recordset.Fields("编号") + ".jpg"
    Picture1.Picture = LoadPicture(zp)
End Sub
Private Sub Bottomcmd_Click()       ' 移动到尾记录
    Adodc1.Recordset.MoveLast
    zp = Adodc1.Recordset.Fields("编号") + ".jpg"
    Picture1.Picture = LoadPicture(zp)
End Sub
Private Sub Upcmd_Click()           ' 移动到上一条记录
    Adodc1.Recordset.MovePrevious
    If Adodc1.Recordset.BOF = True Then
        MsgBox "已到记录头"
    Else
        zp = Adodc1.Recordset.Fields("编号") + ".jpg"
        Picture1.Picture = LoadPicture(zp)
    End If
End Sub
Private Sub Downcmd_Click()         ' 移动到下一条记录
    Adodc1.Recordset.MoveNext
    If Adodc1.Recordset.EOF = True Then
        MsgBox "已到记录尾"
    Else
        zp = Adodc1.Recordset.Fields("编号") + ".jpg"
```

```
            Picture1.Picture = LoadPicture(zp)
        End If
    End Sub
    Private Sub Findcmd_Click()         ' 按查询条件查询
        If Optxm.Value = True Then
            Adodc1.RecordSource = "select * from 人员信息表 where 姓名='" + Trim(Text_xm.Text)
+ "' order by 编号"
            Adodc1.Refresh
        ElseIf Optxb.Value = True Then
            Adodc1.RecordSource = "select * from 人员信息表 where 性别='" + Trim(Text_xb.Text)
+ "' order by 编号"
            Adodc1.Refresh
        ElseIf Optny.Value = True Then
            Adodc1.RecordSource = "select * from 人员信息表 where cdate(出生年月)='" +
Trim(Text_ny.Text) + "' order by 编号"
            Adodc1.Refresh
        ElseIf Optbm.Value = True Then
            Adodc1.RecordSource = "select * from 人员信息表 where 部门='" + Trim(Text_bm.Text)
+ "' order by 编号"
            Adodc1.Refresh
        ElseIf Optwh.Value = True Then
            Adodc1.RecordSource = "select * from 人员信息表 where 文化程度='" +
Trim(Text_wh.Text) + "' order by 编号"
            Adodc1.Refresh
        ElseIf Optzc.Value = True Then
            Adodc1.RecordSource = "select * from 人员信息表 where 职称='" + Trim(Text_zc.Text)
+ "' order by 编号"
            Adodc1.Refresh
        End If
        If Adodc1.Recordset.EOF = True Then
            MsgBox "没有此人！"
        Else
          zp = Adodc1.Recordset.Fields("编号") + ".jpg"
          Picture1.Picture = LoadPicture(zp)
        End If
    End Sub
    Private Sub Cancelcmd_Click()       ' 取消查询
      Text_xm = "":Text_xb = "":Text_ny = "":Text_bm = "":Text_wh = "":Text_zc = ""
      Adodc1.RecordSource = "select * from 人员信息表"
      Adodc1.Recordset.MoveFirst
      zp = Adodc1.Recordset.Fields("编号") + ".jpg"
      Picture1.Picture = LoadPicture(zp)
      Adodc1.Refresh
    End Sub
    Private Sub Findclosecmd_Click()    ' 关闭窗口
       Adodc1.Recordset.Close
       Unload Me
    End Sub
```

六、工资调整界面 gztz.frm 窗体文件事件过程

```
    Private Sub Gzcmd_Click()           ' 工资修改
       Adodc1.Recordset.Update
```

```
        End Sub
        Private Sub Gzclosecmd_Click()       ' 关闭窗口
            Adodc1.Recordset.Close
            Unload Me
        End Sub
```

七、工资核算界面 gzhs.frm 窗体文件事件过程

```
        Private Sub Gzhscmd_Click()          ' 工资核算
            Adodc1.RecordSource = "select 人员信息表.编号,姓名,性别,部门,(请假天数*50+迟到次数*10)
as 应扣款,(基本工资 + 奖金 + 津贴 - 公积金 - 医疗保险 - 会员费) as 应发工资,(应发工资-应扣款) as 实
发工资 from 人员信息表,工资表,考勤表 where 人员信息表.编号=工资表.编号 and 工资表.编号=考勤表.编号
and month(工资表.发放日期)=考勤表.月份"
            Adodc1.Refresh
        End Sub
        Private Sub Gzhsclosecmd_Click()     ' 关闭窗口
            Unload Me
        End Sub
```

八、工资查询界面 gzcx.frm 窗体文件事件过程

```
        Private Sub Gzjscxcmd_Click()        ' 工资计算查询
            If Combo1.Text = "最高工资" Then
                Adodc1.RecordSource = "select max(基本工资) as 最高工资 from 工资表"
                Adodc1.Refresh
            ElseIf Combo1.Text = "最低工资" Then
                Adodc1.RecordSource = "select min(基本工资) as 最低工资 from 工资表"
                Adodc1.Refresh
            ElseIf Combo1.Text = "工资总和" Then
                Adodc1.RecordSource = "select sum(基本工资) as 工资总和 from 工资表"
                Adodc1.Refresh
            ElseIf Combo1.Text = "平均工资" Then
                Adodc1.RecordSource = "select avg(基本工资) as 平均工资 from 工资表"
                Adodc1.Refresh
            ElseIf Combo1.Text = "应发工资" Then
                Adodc1.RecordSource = "select 人员信息表.编号,姓名,性别,部门,发放日期,(基本工资+奖金+
津贴-公积金-医疗保险-会员费) as 应发工资 from 人员信息表,工资表 where 人员信息表.编号=工资表.编号"
                Adodc1.Refresh
            End If
        End Sub
        Private Sub Gzpxcx_Click()           ' 工资排序查询
            If Option_sx = True Then
                Adodc1.RecordSource = "select 人员信息表.编号,姓名,性别,部门,发放日期,基本工资,奖金,津
贴,公积金,医疗保险,会员费 from 人员信息表,工资表 where 人员信息表.编号=工资表.编号 order by 基本工资"
                Adodc1.Refresh
            End If
            If Option_jx = True Then
                Adodc1.RecordSource = "select 人员信息表.编号,姓名,性别,部门,发放日期,基本工资,奖金,津
贴,公积金,医疗保险,会员费 from 人员信息表,工资表 where 人员信息表.编号=工资表.编号 order by 基本工资
desc"
                Adodc1.Refresh
            End If
        End Sub
```

```
Private Sub Gzcxcmd_Click()    ' 工资查询
    If Combo2.Text = "姓名" Then
       Adodc1.RecordSource = "select 人员信息表.编号,姓名,性别,部门,职称,发放日期,基本工资,奖金,津贴,公积金,医疗保险,会员费 from 人员信息表,工资表 where 人员信息表.编号=工资表.编号 and 姓名='" + Trim(Text1.Text) + "'"
       Adodc1.Refresh
    End If
    If Combo2.Text = "性别" Then
       Adodc1.RecordSource = "select 人员信息表.编号,姓名,性别,部门,职称,发放日期,基本工资,奖金,津贴,公积金,医疗保险,会员费 from 人员信息表,工资表 where 人员信息表.编号=工资表.编号 and 性别='" + Trim(Text1.Text) + "'"
       Adodc1.Refresh
    End If
    If Combo2.Text = "部门" Then
       Adodc1.RecordSource = "select 人员信息表.编号,姓名,性别,部门,职称,发放日期,基本工资,奖金,津贴,公积金,医疗保险,会员费 from 人员信息表,工资表 where 人员信息表.编号=工资表.编号 and 部门='" + Trim(Text1.Text) + "'"
       Adodc1.Refresh
    End If
    If Combo2.Text = "职称" Then
       Adodc1.RecordSource = "select 人员信息表.编号,姓名,性别,部门,职称,发放日期,基本工资,奖金,津贴,公积金,医疗保险,会员费 from 人员信息表,工资表 where 人员信息表.编号=工资表.编号 and 职称='" + Trim(Text1.Text) + "'"
       Adodc1.Refresh
    End If
End Sub
Private Sub Gzcxclosecmd_Click()    ' 关闭窗口
    Unload Me
End Sub
```

九、人员考勤界面 rykq.frm 窗体文件事件过程

```
Private Sub Kqeditcmd_Click()         ' 考勤编辑
    Adodc1.Recordset.Update
    Adodc1.Refresh
End Sub
Private Sub Kqcountcmd_Click()        ' 考勤统计
    Adodc1.RecordSource = "select sum(请假天数) as 总请假天数,sum(迟到次数) as 总迟到人次 from 人员信息表,考勤表 where 人员信息表.编号=考勤表.编号"
    Adodc1.Refresh
End Sub
Private Sub Kqclosecmd_Click()        ' 关闭窗口
    Unload Me
End Sub
```

第 14 章
自测综合练习题

14.1 综合练习一

一、选择题

1. 以下能判断是否到达文件尾的函数是（　　）。
   [A] BOF　　　　[B] LOC　　　　[C] LOF　　　　[D] EOF
2. 对于数据控件能将记录指针移动到 Recordset 对象中第一条记录处的方法是（　　）。
   [A] MoveFirst　　[B] MoveNext　　[C] MovePrevious　　[D] MoveLast
3. 刚建立一个新的标准 EXE 工程后，不在工具箱中出现的控件是（　　）。
   [A] 单选按钮　　[B] 图片框　　[C] 通用对话框　　[D] 文本框
4. 要使文本框可输入多行文字，要设置的属性是（　　）。
   [A] ScrollBars　　[B] MultiLine　　[C] Text　　[D] List
5. VB 程序设计采用的编程机制是（　　）。
   [A] 可视化　　[B] 面向对象　　[C] 事件驱动　　[D] 过程结构化
6. 可以在常量的后面加上类型说明符以显示常量的类型，可以表示整型常量的是（　　）。
   [A] %　　　　[B] #　　　　[C] !　　　　[D] $
7. 图像框或图片框中显示的图形，由对象的（　　）属性值决定。
   [A] Picture　　[B] Image　　[C] Icon　　[D] MouseIcon
8. 当拖动滚动条时，将触发滚动条的（　　）事件。
   [A] Move　　[B] Change　　[C] Scroll　　[D] GotFocus
9. 若要向列表框中添加内容，可使用的方法是（　　）。
   [A] Add　　　[B] Remove　　[C] Clear　　[D] AddItem
10. 窗体文件的扩展名为（　　）。
    [A] vbp　　　[B] bas　　　[C] exe　　　[D] frm
11. 下面将 x 定义为单精度变量的是（　　）。
    [A] x$　　　　[B] x!　　　　[C] x#　　　　[D] x%
12. 窗体标题栏的显示内容由窗体的（　　）属性决定。
    [A] Name　　[B] Caption　　[C] BackColor　　[D] Enabled
13. 调用公用对话框 Commondialog1 中，"打开"对话框的方法是（　　）。

[A] ShowSave [B] ShowPrinter [C] ShowHelp [D] Showopen

14. VB 表达式 Sqr(a)^3*2-5 中优先进行运算的是（ ）。

    [A] Sqr 函数 [B] - [C] ^ [D] *

15. 若要从顺序文件中读取内容，则设置其打开方式为（ ）。

    [A] Append [B] Input [C] Output [D] RW

16. 设置标签背景颜色，应设置（ ）属性来实现。

    [A] BackColor [B] ForeColor [C] Fontcolor [D] Fontname

17. 设 a="Visual Basic"，下面使 b="Basic" 的语句是（ ）。

    [A] b=Left(a,8,12) [B] b=Mid(a,8,5) [C] b=Right(a,5,5) [D] b=Left(a,8,5)

18. 写在一行上的多条语句，应以（ ）为分隔符。

    [A] 分号 [B] 逗号 [C] 冒号 [D] 空格

19. 在窗体上画一个名称为 Timer1 的计时器控件，要求每隔 0.5s 发生一次计时器事件，则以下正确的属性设置语句是（ ）。

    [A] Timer1.InterVal=0.5 [B] Timer1.Interval=500
    [C] Timer1.Interval=50 [D] Timer1.Interval=5

20. 能够获得一个文本框中被选取文本的内容的属性是（ ）。

    [A] Text [B] SelLength [C] Seltext [D] SelStart

21. 若要获得滚动条的当前值，可通过访问其（ ）属性来实现。

    [A] Text [B] Min [C] Max [D] Value

22. 确定窗体控件启动位置的属性是（ ）。

    [A] Width 和 Height [B] Width 或 Height
    [C] X,Y [D] Top 和 Left

23. 将数值型数据转换成字符型数据的函数为（ ）。

    [A] Val() [B] Str() [C] Trim() [D] Len()

24. 下列赋值语句中错误的是（ ）。

    [A] x = x+1 [B] x = x+y [C] x+y = x [D] x = 4>6

25. 数据控件集 Recordset 添加记录的方法为（ ）。

    [A] Move [B] Addnew [C] Edit [D] Update

26. 以下控件不能作为容器的是（ ）。

    [A] 窗体 [B] 图片框 [C] 图像框 [D] 框架

27. 在 VB 集成开发环境中要结束一个正在运行的工程，可单击工具栏上的一个按钮，这个按钮是（ ）。

    [A] ▶ [B] ❚❚ [C] ■ [D] ↺

28. VB 提供的选择框（CheckBox）可具有的功能是（ ）。

    [A] 多重选择 [B] 单一选择 [C] 多项选择 [D] 选择一次

29. 执行语句 s=Len(Mid("abcdefghijklm",1,6)) 后，s 的值是（ ）。

    [A] abcdef [B] ghijkl [C] 6 [D] 12

30. 以下关系表达式中，其值为 False 的是（ ）。

    [A] "LMN">"LmN" [B] "the"<>"they"
    [C] "VISUAL"=UCase("Visual") [D] "Integer">"Int"

二、填空题

1. VB 的菜单可分为_____和快捷（弹出）菜单两种。
2. 将数学表达式 $\frac{4}{3}\sqrt{a^2+b^2}$ 改写为 VB 表达式为_____。
3. 表达式 14 mod 3 的值为_____。
4. 声明全局变量一般使用_____关键字。
5. 常量是指在程序运行过程中其值_____的量。
6. VB 中的日期型常量用_____引起来。
7. 表达 x 是 5 的倍数或是 9 的倍数的逻辑表达式为_____。
8. 若要将窗体 form1 隐藏起来，可以调用其方法_____来实现。
9. VB 中的注释语句以_____开头。
10. 逻辑常量值为 True 或_____。

三、判断题

1. 在窗体标题栏显示的图标应使用 Picture 属性。（   ）
2. 窗体的高度和宽度可用 Height 和 Width 属性进行设置。（   ）
3. 运行窗体时会触发 Load 事件。（   ）
4. 注释语句可以写在命令的后面，注释的内容要以 Rem 或单引号'开头。（   ）
5. C1 是公用对话框控件，命令 C1.ShowFont 可以打开字体对话框。（   ）
6. 当用户单击工具栏上按钮时引发的事件是 Click 事件。（   ）
7. 图片框能用 Move 方法移动。（   ）
8. 启动计时器的属性是 Interval。（   ）
9. 列表框中的数据只能在设计时用属性窗口添加，不能在程序运行时用命令添加。（   ）
10. 单选按钮被选中时其 Value 值为 True，否则为 False。（   ）

四、程序改错题

1. 请根据下列描述编写购物优惠程序。某商场为了加速促成商品流通，采用购物打折的优惠办法：每位顾客一次购物（1）在 100 元及以上者，按九五折优惠；（2）在 200 元及以上者，按九折优惠；（3）300 元及以上者，按八折优惠；（4）500 元及以上者按七折优惠。

```
Option Explicit
Private Sub Command1_Click()
   Dim x As Single, y As Single
   x = Val(Text1.Text)
'**********FOUND**********
   If x > 100 Then
     y = x
   ElseIf x < 200 Then
     y = 0.95 * x
   ElseIf x < 300 Then
     y = 0.9 * x
   ElseIf x < 500 Then
     y = 0.8 * x
   ElseIf x >= 500 Then
     y = 0.7 * x
'**********FOUND**********
   Else
```

```
'**********FOUND**********
    Text2.Text = x
End Sub
```

2. 下边是一个小动画程序，在窗体上放一个标签 Label1，每过 1s 标签 Label1 的背景颜色由红到蓝，由蓝到绿，再由绿到红循环变化，并自动修改标签 Label1 的 Left，Top 值使其从左上角沿窗体的对角线移动到窗体的右下角，如此往复从而实现动画。

```
Option Explicit
Private flag As Integer
Private Sub Form_click()
    Timer1.Enabled = True
End Sub

Private Sub Form_Load()
    Timer1.Interval = 1000
    Label1.Left = 0
    Label1.Top = 0
    Label1.BackColor = vbBlack
    Timer1.Enabled = False
End Sub
Private Sub Timer1_Timer()
    If Label1.Left < Left + Width And Label1.Top <= Top + Height Then
        Label1.Left = Label1.Left + 100 * Width / Height
        Label1.Top = Label1.Top + 100
    Else
        Label1.Left = Left
        Label1.Top = Top
    End If
    If flag = 0 Then
        Label1.BackColor = vbBlue
        '**********FOUND**********
        flag = 0
    ElseIf flag = 1 Then
        Label1.BackColor = vbGreen
        '**********FOUND**********
        flag = 1
    Else
        Label1.BackColor = vbRed
        '**********FOUND**********
        flag = 2
    End If
End Sub
```

五、程序填空题

1. 素数是一个只能被 1 和自身整除的正整数。用鼠标单击事件可以完成如下程序设计，将 100～300 的素数打印出来。

```
PUBLIC T AS Boolean
Dim I as integer
Dim J as integer
Sub qssgc(n as integer)
    T=false
    J = int(sqr( n))
```

```
      For I= 2 TO J
      '**********SPACE**********
        If 【?】 then
          Exit sub
        End if
      Next I
      T= true
   End sub
   Private sub  command1_click()
     Dim n as integer
     '**********SPACE**********
      For n= 【?】
     '**********SPACE**********
      【?】
        If t then print spc(1) ; n ;
      Next n
    End sub
```

2. 用顺序文件的方式读取文件"C：\AUTOEXEC.BAT"的内容，并显示在窗体上的文本框中。

```
Private Sub Form_Activate()
   Dim textline As String
   '**********SPACE**********
   open 【?】
   '**********SPACE**********
   Do While 【?】
     Input  #1, textline
     text1.Text =text1.text+textline
   Loop
   Close  #1
End Sub
```

3. 在下列事件过程中，如果选中复选框 1，则文本变成斜体，如果选中复选框 2，则 Text1 的背景色变成蓝色。

```
Private Sub Check1_Click( )
   If Check1.Value=1 Then
   '**********SPACE**********
     Text1. FontItalic= 【?】
   Else
     Text1.FontItalic=False
   End If
End Sub
Private Sub Check2_Click( )
   If Check2.Value=1 Then
     Text1. BackColor =vbBlue
   Else
     Text1. BackColor =vbBlack
   End If
End Sub
```

六. 编程题

编写函数 fun，函数的功能是：求从 m 到 n 的奇数的乘积并显示，如：m 为 2，n 为 5 时，显示"15"。'存储连乘的乘积的变量必须为 Product,'要求使用 For 语句来实现。

```
Private Function fun(m As Integer, n As Integer) As Long
    Dim Product As Double, t As Integer
    If m > n Then t = m: m = n: n = t
    '**********Program**********

    '********** End **********
End Function

Private Sub Form_Load()
    Show
    Print fun(5, 2)
    NJIT_VB
End Sub

Private Sub NJIT_VB()
    Dim i As Integer
    Dim a(10) As String
    Dim fIn As Integer
    Dim fOut As Integer
    fIn = FreeFile
    Open App.Path & "\in.dat" For Input As #fIn
    fOut = FreeFile
    Open App.Path & "\out.dat" For Output As #fOut
    For i = 1 To 10 Step 2
      Line Input #fIn, a(i)
      Line Input #fIn, a(i + 1)
      Print #fOut, Trim(Str(fun(Val(a(i)), Val(a(i + 1)))))
    Next
    Close #fIn
    Close #fOut
End Sub
```

14.2 综合练习二

一、选择题
1. 如果一个变量未经定义就直接使用,则该变量的类型为(   )。
   [A] Integer　　　　[B] Byte　　　　[C] Boolean　　　　[D] Variant
2. VB 模块文件的扩展名是(   )。
   [A] .bas　　　　　[B] .cls　　　　 [C] .frm　　　　　 [D] .vbp
3. 确定一个控件宽度和高度的属性是(   )。
   [A] Width 和 Height　　　　　　　　[B] Width 或 Height
   [C] Top 和 Left　　　　　　　　　　[D] Top 或 Left
4. 下列正确的赋值语句是(   )。
   [A] m+1=m　　　 [B] m=m−1　　　[C] 3m=n+x　　　 [D] −m=n+1
5. 表达式 10+10Mod2^3+3 的值是(   )。

[A] 7　　　　　　[B] 9　　　　　　[C] 15　　　　　[D] 10

6. 设窗体上有一列表框 List1，且其中含有若干列表项。则以下能表示当前被选中的列表项索引的是（　　）。

[A] List1.List　　[B] List1.ListIndex　　[C] List1.Index　　[D] List1.Text

7. Dim A(5, 5) As Integer 定义的数组包含的元素个数是（　　）。

[A] 25　　　　　[B] 动态变化　　[C] 30　　　　　[D] 36

8. 控件的获得控制焦点事件为（　　）。

[A] Refresh　　[B] SetFocus　　[C] GotFocus　　[D] Value

9. 若要使图片框的大小自动与所显示的图片相适应，则可通过设置（　　）属性的值为 True 来实现。

[A] AutoSize　　[B] Alignment　　[C] Appearance　　[D] Visible

10. 如果准备向随机文件中写入数据，正确的语句是（　　）。

[A] Print # 1,rec　　[B] Write # 1,rec　　[C] Put # 1,rec　　[D] Get # 1,rec

11. 要使文本框拥有滚动条，要设置的属性是（　　）。

[A] ScrollBars　　[B] MultiLine　　[C] Text　　[D] List

12. 在过程定义中用（　　）表示形参的传值。

[A] Var　　　　　[B] ByDef　　　[C] ByVal　　　　[D] Value

13. 设 a=10,b=5,c=1，执行语句 Print a>b<c 后，窗体上显示的是（　　）。

[A] True　　　　[B] False　　　　[C] 1　　　　　　[D] 出错信息

14. 以下能判断打开文件大小的函数是（　　）。

[A] BOF　　　　[B] LOC　　　　[C] LOF　　　　[D] EOF

15. 图像框或图片框中显示的图像，由对象的（　　）属性值决定。

[A] Picture　　　[B] Image　　　[C] Icon　　　　[D] MouseIcon

16. 数学关系 8≤y<28 表示成正确的 VB 表达式为（　　）。

[A] 8<=y<28

[B] 8<=y AND<28

[C] 8<=y AND y<28

[D] 8<=y OR y<28

17. 若要设置滚动条的最小值，可通过设置其（　　）属性来实现。

[A] Text　　　　[B] Min　　　　[C] Max　　　　[D] Value

18. 当滚动条的滚动块移动后，将触发滚动条的（　　）事件。

[A] Move　　　　[B] Change　　　[C] Scroll　　　[D] GotFocus

19. 若要获知当前列表框列表项的数目，可通过访问（　　）属性来实现。

[A] List　　　　[B] ListIndex　　[C] ListCount　　[D] Text

20. 若要删除列表框中某一项内容，可使用的方法是（　　）。

[A] Add　　　　[B] RemoveItem　　[C] Clear　　　　[D] AddItem

21. 若要以程序代码方式设置在窗体中显示文本的字体，则可用窗体对象的（　　）属性来实现。

[A] FontName　　[B] Font　　　[C] FontSize　　　[D] FontBold

22. 设窗体上有一个文本框，名称为 text1，程序运行后，要求该文本框不可见，以下能实现该操作的语句是（　　）。

[A] Text1.MaxLength=0　　　　[B] Text1.Enabled=False

[C] Text1.Visible=False　　　　　　[D] Text1.Width=0

23. 要在窗体 Form1 内显示"新年快乐"，使用的语句是（　　）。

　　[A] Form. print "新年快乐"　　　　[B] Form1.Title="VisualBasic 窗体"

　　[C] Form1.Caption="新年快乐"　　[D] Form1.print "新年快乐"

24. 设菜单中有一个菜单项为"Open"。若要为该菜单命令设置访问键，即按下 Alt 及字母 O 时，能够执行"Open"命令，则在菜单编辑器中设置"Open"命令的方式是（　　）。

　　[A] 把 Caption 属性设置为&Open　　[B] 把 Caption 属性设置为 O&pen

　　[C] 把 Name 属性设置为&Open　　　[D] 把 Name 属性设置为 O&pen

25. 用 InputBox 函数设计的对话框，其功能是（　　）。

　　[A] 只能接收用户输入的数据，但不会返回任何信息

　　[B] 能接收用户输入的数据，并能返回用户输入的信息

　　[C] 既能用于接收用户输入的信息，又能用于输出信息

　　[D] 专门用于输出信息

26. 在窗体上画一个名称为 Timer1 的计时器控件，要求每隔 5s 发生一次计时器事件，则以下正确的属性设置语句是（　　）。

　　[A] Timer1.Interval=0.5　　　　　[B] Timer1.Interval=5

　　[C] Timer1.Interval=5000　　　　[D] Timer1.Interval=500

27. 图像框有一个属性，可以自动调整图形的大小，以适应图像框的尺寸，这个属性是（　　）。

　　[A] Autosize　　[B] Stretch　　[C] AutoRedraw　　[D] Appearance

28. 以下叙述中错误的是（　　）。

　　[A] 事件过程是响应特定事件的一段程序

　　[B] 不同的对象可以具有相同名称的方法

　　[C] 对象的方法是执行指定操作的过程

　　[D] 对象事件的名称可以由编程者指定

29. 以下合法的 VB 标识符是（　　）。

　　[A] Integer　　[B] ab-c　　[C] 56abc　　[D] ab_c

30. 当一个单选按钮被选中时，它的 Value 属性的值是（　　）。

　　[A] True　　[B] False　　[C] 1　　[D] 0

二、填空题

1. 要将隐藏的窗体 form1 显示出来，可使用语句_____。

2. VB 的菜单可分为_____和_____两种。

3. 若要以输入模式打开 D 盘根目录下 ee.txt 顺序文件，打开的语句为：_____ As #1。

4. 逻辑常量值为 False 或_____。

5. 将图片框 Picture1 中的图像文件删除的语句是 Picture1.Picture=_____。

6. 双精度数用字母_____将尾数与指数分开。

7. Print 以标准格式在窗体上输出数据，数据之间应使用_____。

8. 清除窗体 Form1 上的图形和文字应使用命令_____。

9. 函数 Sqr(x)的功能是_____。

三、判断题

1. 要在 VB 程序中加注释，可使用 REM 或单引号。（   ）
2. 一个应用程序中只能创建一个窗体。（   ）
3. VB 中的数值可以用十六进制或八进制表示，八进制数的开头符是&H。（   ）
4. VB 中的过程不能嵌套定义，但能嵌套调用。（   ）
5. 能够获得一个文本框中被选取文本的长度的属性是 SelStart。（   ）
6. VB 中每个菜单项除了 Click 事件之外，不可以响应其他事件。（   ）
7. 标签控件没有 Caption 属性。（   ）
8. 任何控件都有 Name 属性。（   ）
9. Right 函数的作用是取右子串。（   ）
10. PictureBox 不可作为其他控件的容器。（   ）

四、程序改错题

1. 以下程序用于建立一个五行五列的矩阵，使其两条对角线上数字为 1，其余位置为 0。

```
Option Explicit
Private Sub Form_Click()
Dim x(5, 5), n As Integer, m As Integer
  For n = 1 To 5
    For m = 1 To 5
'**********FOUND**********
      If n = m Or m + n = 4 Then
        x(n, m) = 1
      Else
        x(n, m) = 0
      End If
    Next m
  Next n
'**********FOUND**********
  For n = 1 To 3
    For m = 1 To 5
'**********FOUND**********
      Print x(m, n)
    Next m
    Print
  Next n
End Sub
```

2. 密码判断程序，如果密码为 12345 则显示"恭喜，密码正确"，否则显示"很遗憾，密码错误"，要求文本框中只允许输入数字。

```
Option Explicit
Private Sub Command1_Click()
  Dim strPws As String
  strPws = Trim(Text1.Text)
'**********FOUND**********
  If Len(strPws) <> 0 Then Exit Sub
  If strPws = "12345" Then
'**********FOUND**********
    MsgBox "恭喜，密码正确", 验证
  Else
```

```
        MsgBox "很遗憾,密码错误",,"验证"
    End If
End Sub
Private Sub Text1_KeyPress(KeyAscii As Integer)
    '**********FOUND**********
    If Not (KeyAscii >= 49 And KeyAscii <= 57) Then
        KeyAscii = 0
    End If
End Sub
```

五、程序填空题

1. 在窗体上有一个过程函数,然后编写如下事件程序,该过程的功能是用选择交换法将 10 个数排成升序,请在空白处填入适当的语句。

```
Sub SORT()
    Dim a(1 To 10)
    For i=1 To 10
        a(i) =Val(InputBox("", "", 0) )
    Next i
    '**********SPACE**********
    For i=【?】
        k=i
    '**********SPACE**********
        For j=【?】
    '**********SPACE**********
            If a(k) > a(j) Then 【?】
            If k <> i Then
                b=a(k)
                a(k) =a(i)
                a(i) =b
            End If
        Next j
    Next i
    For k=1 To 10
        Print a(k)
    Next k
End Sub
```

2. 在窗体上有一个"背景色变换"按钮和一个"结束"按钮。单击"背景色变换"按钮,背景色 变为红色;再单击,背景色变为绿色;再单击,背景色变为蓝色;再单击背景色变为红色…如此循环。请填空。

```
Private Sub cmdChange_Click()
    If Mark=0 Then
    '**********SPACE**********
        【?】
        Mark=1
    '**********SPACE**********
    ElseIf【?】Then
        Form1. BackColor=vbGreen
        Mark=2
    ElseIf Mark=2 Then
```

```
        Form1.BackColor=vbBlue
    '**********SPACE**********
        【?】
    End If
End Sub
```

六、编程题

应用冒泡法对数组 A 按升序排列。基本思想：**(将相邻两个数比较，小的调到前头)**

（1）有 n 个数（存放在数组 a(n)中），第一趟将每相邻两个数比较，小的调到前头，经 n-1 次两两相邻比较后，最大的数已"沉底"，放在最后一个位置，小数上升"浮起"；

（2）第二趟对余下的 n-1 个数（最大的数已"沉底"）按上法比较，经 n-2 次两两相邻比较后得次大的数；

（3）依次类推，n 个数共进行 n-1 趟比较，在第 j 趟中要进行 n-j 次两两比较。

```
Option Explicit
Private Sub Sort(ByRef a() As Integer, n As Integer)
'********** Program *********



'********** End *************
End Sub
Private Sub Form_Load()
Show
    Dim i As Integer
    Dim arr(10) As Integer
    For i = 1 To 10
        arr(i) = Int(10 * Rnd + 1)
    Next
    Sort arr, 10
    For i = 1 To 10
        Print arr(i)
    Next
    WWJT
End Sub
Private Sub WWJT()
    Dim i As Integer
    Dim s As String
    Dim l As Long
    Dim d As Double
    Dim a(10) As String
    Dim b(10) As Integer
    Dim fIn As Integer
    Dim fOut As Integer
    fIn = FreeFile
    Open App.Path & "\in.dat" For Input As #fIn
    fOut = FreeFile
    Open App.Path & "\out.dat" For Output As #fOut
    For i = 1 To 10
        Line Input #fIn, a(i)
        l = Val(a(i))
```

```
      b(i) = 1
   Next
   Sort b, 10
   For i = 1 To 10
      Print #fOut, b(i)
   Next
   Close #fIn
   Close #fOut
End Sub
```

参考文献

[1] 肖红. VB 语言程序设计实验指导与习题解答[M]. 北京：人民邮电出版社，2010.
[2] 王栋. Visual Basic 程序设计实用教程[M]. 北京：清华大学出版社，2013.
[3] 邹丽明. Visual Basic 6.0 程序设计与实训[M]. 北京：电子工业出版社，2008.
[4] 范晓平. Visual Basic 软件开发姓名实训[M]. 北京：海洋出版社，2006.